DFG

The MAK-Collection for Occupational Health and Safety

Part IV: Biomonitoring Methods

Part I: MAK Value Documentations
ISSN 1860-496X

Part II: BAT Value Documentations
ISSN 1860-4978

Part III: Air Monitoring Methods
ISSN 1860-4986

Part IV: Biomonitoring Methods
ISSN 1860-4994

The MAK-Collection online
www.mak-collection.com

DFG *Deutsche Forschungsgemeinschaft*

The MAK-Collection for Occupational Health and Safety

Part IV: Biomonitoring Methods
Volume 13

Edited by Thomas Göen
Working Group Analyses in Biological Materials

Commission for the Investigation of Health Hazards of Chemical Compounds in the Work Area
(Chairwoman: Andrea Hartwig)

WILEY-VCH Verlag GmbH & Co. KGaA

Prof. Dr. Andrea Hartwig
Karlsruher Institut für Technologie (KIT)
Institut für angewandte Biowissenschaften
Abteilung Lebensmittelchemie und Toxikologie
Adenauerring 20
76131 Karlsruhe
Germany

Prof. Dr. Thomas Göen
Institut und Poliklinik für Arbeits-,
Sozial- und Umweltmedizin
Friedrich-Alexander-Universität Erlangen-Nürnberg
Schillerstr. 25/29
91054 Erlangen
Germany

Vol. 1–9 were published under the title "Analyses of Hazardous Substances in Biological Materials"
(ISSN 0179-7247)

All books published by Wiley-VCH are carefully produced. Nevertheless, authors, editors, and publisher do not warrant the information contained in these books, including this book, to be free of errors. Readers are advised to keep in mind that statements, data, illustrations, procedural details or other items may inadvertently be inaccurate.

Library of Congress Card No.: applied for

British Library Cataloging-inPublication Data:
A catalogue record for this book is available from the British Library.

Bibliographic information published by the Deutsche Bibliothek
The Deutsche Bibliothek lists this publication in the Deutsche Nationalbibliografie; detailed bibliographic data is available on the Internet at http://dnb.d-nb.de.

© 2013 WILEY-VCH Verlag GmbH & Co. KGaA, 69469 Weinheim, Germany

All rights reserved (including those of translation into other languages). No part of this book may be reproduced in any form by photoprinting, microfilm, or any other means nor transmitted or translated into a machine language without written permission from the publishers. Registered names, trademarks, etc. used in this book, even when not specifically marked as such, are not to be considered unprotected by law.

ISSN: 1860-4994
ISBN: 978-3-527-33438-4

Cover Design Grafik-Design Schulz, Fußgönheim
Typesetting Beltz Bad Langensalza GmbH, Bad Langensalza
Printing and Binding betz-druck GmbH, Darmstadt, Germany

Printed on acid-free paper

Preface

Over the last decades, biological monitoring has proved to be an effective instrument in the assessment of individual risks when dealing with hazardous substances. At a very early stage, the DFG Senate Commission for the Investigation of Health Hazards of Chemical Compounds in the Work Area (DFG Commission) recognised the possibilities offered by human biomonitoring. These are in particular: the determination of chemical substances and of their metabolites in human body fluids for preventive examinations in occupational medicine. Therefore, the DFG established the working group "Analyses in Biological Materials" in the early 1970s. The task of this group was to promote reliable analytical procedures as a basis on which the performance of human biomonitoring could be realized. This decision of the DFG Commission and its consistent implementation in practice has considerably contributed to the fact that biological monitoring is nowadays established – not only in Germany but also on an international basis – in the assessment of health hazards in occupational and environmental medicine.

Right from the beginning, it has been the idea of the working group "Analyses in Biological Materials" to assess the reliability of analytical procedures on the basis of objective criteria. The founders of the working group recognised the fact that the best way to obtain a clear statement on the reliability of an analytical procedure is by performing a thorough examination, i.e. via reproduction of the method in at least one laboratory by an independent expert. Consequently, the scientific working group consists of experts who possess the necessary experience in performing and validating biomonitoring procedures and are in a position to reproduce and examine external procedures in their own laboratories. Only this kind of practical examination of the analytical methods makes it possible to assess whether a particular operating procedure is sufficiently detailed and clearly formulated to actually reproduce it without problems in any other laboratory. Moreover, the most important data on reliability such as precision, accuracy and limits of detection are determined by the examiner in a separate validation process and compared with the data given by the author. The examined biomonitoring procedures published in this series by the working group are consequently not only verified as concerns their reliability criteria, but are also available in the form of detailed and reproducible standard operating procedures. Thus, they can be established without problems in laboratories possessing the corresponding equipment.

With the present volume this series of methods is extended by 13 additonal analytical procedures for the quantitation of biomarkers of important chemicals in the work area such as cyclohexane, N,N-dimethylformamide, epichlorohydrin, epoxypropane, isopropyl benzene (cumene) and tetrahydrofuran. In addition, the present volume also describes procedures that allow the determination of background exposure levels in the general population. These include the methods for the determination of bisphenol A in urine, of the phytoestrogens genistein, daidzein and equol in urine, of the metabolites of pyrethroid-based pesticides in urine and of the mercapturic acids of acrylamide, acrolein and epoxypropane in urine. The present volume contains procedures for quite a number of biomarkers that have been included for the first time in the series of methods issued by the DFG Commission. Thus, some existing gaps have now been filled.

I wish to thank all members, guests and ad-hoc experts who have, thanks to their excellent competence and high level of commitment on a voluntary basis, made this publication possible. Moreover, I would like to extend special thanks to the *Deutsche Forschungsgemeinschaft* for their efficient support of the Senate Commission's work.

Prof. Dr. Andrea Hartwig
Chair of the DFG Senate Commission for the Investigation of Health Hazards of Chemical Compounds in the Work Area

Foreword

The 13th volume of the English publication series "MAK Collection, Part IV: Biomonitoring Methods" is being published simultaneously with the 20th Supplement Issue of the loose-leaf collection „*Analytische Methoden zur Prüfung gesundheitsschädlicher Arbeitsstoffe, Band 2: Analysen in biologischem Material*". Both versions of the method collections are now complemented by 13 methods for the determination of exposure to hazardous substances in human biological material, which have been positively proved by the working group "Analyses in Biological Materials" of the DFG Senate Commission for the Investigation of Health Hazards of Chemical Compounds in the Work Area (DFG Commission). Thus, in the volumes of the English edition, a total of 152 analytical methods are now available. In the present German edition, 216 proved methods for the determination of over 300 parameters have been published, which can be used in the context of biological exposure and effect monitoring or for the determination of susceptibility parameters.

The present 13th volume fills some gaps in the methods collection. Of great significance in this respect is the inclusion of procedures for some parameters with already existing limit values or reference values in biological materials but for which proved analytical methods were not previously available in the method collection. This includes the method for the determination of cyclohexanol and 1,2-cyclohexanediol in urine as parameters of exposure to cyclohexane and cyclohexanone, for which a BAT value (*Biologischer Arbeitsstoff-Toleranzwert*; biological limit value) and exposure equivalents for carcinogenic substances (EKA, *Expositionsäquivalente für krebserzeugende Arbeitsstoffe*) are available. For both of these parameters, Biological Exposure Indices (BEI) of the American Conference of Governmental Industrial Hygienists (ACGIH) have now also become available for a biomonitoring of cyclohexanone. Moreover, a procedure for the determination of 2-phenyl-2-propanol in urine as biomarker of exposure to isopropyl benzene (cumene) was now added to the method collection, too. With this method, a monitoring of the BAT value for 2-phenyl-2-propanol is made possible, that was established by the DFG Commission in 2000. The procedure for the determination of tetrahydrofuran (THF) in urine, which was written as an addendum to the DFG method "Alcoholes and Ketones", also fills such a gap, as reference values such as BAT and BEI are now available for THF. Furthermore, the present issue covers, for the first time, two procedures by means of which N-acetyl-S-(N-methylcarbamoyl)-cysteine (AMCC) in the urine can be quantified. A BEI value for AMCC as biomarker of exposure to N,N-dimethylformamide has been established by the ACGIH.

In addition to this, methods for detecting exposure to alkylating hazardous substances can be found in the present volume. These include the methods for the determination of the haemoglobin adducts of glycidol and epichlorohydrin as well as those for the determination of the mercapturic acids of acrolein, acrylamide, ethylene oxide and propylene oxide (epoxypropane).

Many of the procedures published here also enable a quantitation of the biomarkers in the concentration range of environmental exposure or a determination of the background level in the general population. In this context, for example, the new methods for the determination of total arsenic in urine, for the determination of pyrethroid metabolites in urine as well as for the determination of bisphenol A and the phytoestrogens genistein, daidzein and equol in urine should be noted.

Without the special personal commitment of the members, guests and ad hoc experts of the working group who have not only developed and examined the methods on a voluntary basis but who have also provided analytical instruments and further laboratory equipment for this work, it would not have been possible to provide such a method collection. To them and the institutions in which they completed their tasks I wish to express my warmest thanks. I would also like to thank Dr. Elisabeth Eckert and Dr. Anja Schäferhenrich, who supported our endeavors in the scientific secretariat of the working group, as well as Dr. Rudolf Schwabe of the scientific secretariat of the DFG Commission. And, quite naturally, my special thanks go to the DFG and here very particularly to Dr. Ingrid Ohlert, Dr. Armin Krawisch and Sandra Nitz. Their valuable support enables the efficient work of the AibM-working group.

Prof. Dr. Thomas Göen
Chairman of the working group "Analyses in Biological Materials" of the DFG Senate Commission for the Investigation of Health Hazards of Chemical Compounds in the Work Area.

Contents

Contents of Volumes 1–13 XI

The Working Group "Analyses in Biological Materials" of the DFG Senate Commission for the Investigation of Health Hazards of Chemical Compounds in the Work Area XXXVI

Organisation – The working group's tasks – The work flow within the working group – Publications by the working group – Withdrawal of methods

Terms and symbols used XXXVIII

Analytical Methods

N-Acetyl-S-(N-methylcarbamoyl)-cysteine (AMCC), N-hydroxymethyl-N-methylformamide (HMMF) and N-methylformamide (NMF) in urine	3
Arsenic (total) in urine.	21
Bisphenol A, genistein, daidzein and equol in urine	37
N-(3-Chloro-2-hydroxypropyl)-valine in blood as haemoglobin adduct of epichlorohydrin	63
Cyclohexanol, 1,2-cyclohexanediol and 1,4-cyclohexanediol in urine. . . .	85
N-(2,3-Dihydroxypropyl)-valine in blood as haemoglobin adduct of glycidol	101
Mercapturic acids (N-acetyl-S-(2-carbamoylethyl)-L-cysteine, N acetyl-S-(2-hydroxyethyl)-L-cysteine, N-acetyl-S-(3-hydroxypropyl)-L-cysteine, N-acetyl-S-(2-hydroxypropyl)-L-cysteine, N acetyl-S-(N-methylcarbamoyl)-L-cysteine) in urine	123
2-Phenyl-2-propanol in urine	163
Pyrethrum and pyrethroid metabolites (after solid phase extraction) in urine	179
Pyrethrum and pyrethroid metabolites (after liquid liquid extraction) in urine	215
Selenium in serum.	249
Tetrahydrofuran (THF) in urine (Addendum to the DFG-method "Alcohols and Ketones").	265
Thiocyanate in plasma and saliva	277

Members, Guests and ad-hoc Experts of the Working Group Analyses in Biological Materials of the Senate Commission of the Deutsche Forschungsgemeinschaft for the Investigation of Health Hazards of Chemical Compounds in the Work Area 293

Contents of Volumes 1–13

Substance	Vol.	Page
AAMA	see mercapturic acids	
Acetone	see alcohols and ketones	
Acetonitrile	see thiocyanate	
Acetylcholinesterase (AchE; acetylcholine-acetylhydrolase EC 3.1.1.7) in erythrocytes and cholinesterase (ChE: acylcholin-acylhydrolase EC 3.1.1.8) in plasma	3	45
Acetylcholine-acetylhydrolase	see acetylcholinesterase and cholinesterase	
N-Acetyl-S-(2-carbonamidethyl)-cysteine	see mercapturic acids	
N-Acetyl-S-(N-methylcarbamoyl)-cysteine (AMCC), N-hydroxymethyl-N-methylformamide (HMMF) and N-methylformamide (NMF) in urine	13	3
N-Acetyltransferase 2 (genotyping)	9	135
N-Acetyltransferase 2 (phenotyping)	9	165
AChE	see acetylcholinesterase and cholinesterase	
Acrolein	see mercapturic acids	
Acrylamide	see N-(2-carbamoylethyl)valine	
Acrylamide	see mercapturic acids	
Acrylonitrile	see N-2-cyanoethylvaline, N-2-hydroxyethylvaline, N-methylvaline	
Acrylonitrile	see thiocyanate	
Acylcholin-acylhydrolase	see acetylcholinesterase and cholinesterase	
ADBI	see polycyclic musk compounds	
AHDI	see polycyclic musk compounds	
AHTN	see polycyclic musk compounds	
Alcohols and ketones (acetone; 1-butanol; 2-butanol; 2-butanone; ethanol; 2-hexanone; methanol; 2-methyl-1-propanol; 4-methyl-2-pentanone; 1-propanol, 2-propanol) in blood and urine	5	1

The MAK-Collection Part IV: Biomonitoring Methods, Vol. 13
DFG, Deutsche Forschungsgemeinschaft
© 2013 Wiley-VCH Verlag GmbH & Co. KGaA. Published 2013 by Wiley-VCH Verlag GmbH & Co. KGaA

Substance	Vol.	Page
Alkoxycarboxylic acids in urine as metabolites of glycol ethers with a primary alcohol group	10	55
Allethrin	see pyrethrum and pyrethroid metabolites	
Aluminium, chromium, cobalt, copper, manganese, molybdenum, nickel, vanadium in urine	7	73
Aluminium in plasma	6	47
AMCC	see mercapturic acids	
AMCC	see N-Acetyl-S-(N-methylcarbamoyl)-cysteine	
Aminodinitrotoluenes in urine as metabolites of trinitrotoluene	10	81
2-Amino-4,6-dinitrotoluene	see aminodinitrotoluenes	
4-Amino-2,6-dinitrotoluene	see aminodinitrotoluenes	
4-Aminodiphenyl	see aromatic amines	
4-Aminodiphenyl	see haemoglobin adducts of aromatic amines	
Aminotoluenes	see haemoglobin adducts of aromatic amines	
Amitrole (3-amino-1,2,4-triazole) in urine	6	63
Aniline	see aromatic amines	
Aniline	see haemoglobin adducts of aromatic amines	
o-Anisidine	see aromatic amines	
Anthracycline cytostatic agents (doxorubicin, epirubicin, daunorubicin, idarubicin) in urine	7	119
Antimony in blood and urine	2	31
Antimony in urine	4	51
Antimony	see ICP-MS collective method	
Application of the ICP-MS for biological monitoring	6	1
Aromatic alcohols	see phenols and aromatic alcohols	

Substance	Vol.	Page
Aromatic amines in urine (1-naphthylamine; 2-naphthylamine; 4,4′-methylene-bis(2-chloroaniline); 3,3-dichlorobenzidine)	1	17
Aromatic amines (aniline; o-toluidine; m-toluidine; p-toluidine; 2,4- and 2,6-toluylenediamine; 4-aminodiphenyl; 4,4′-diaminodiphenylmethane) in urine, plasma and erythrocytes	4	67
Aromatic carboxylic acids in urine (phenylglyoxylic acid; mandelic acid; hippuric acid; o-methylhippuric acid, m-/p-methylhippuric acids; benzoic acid)	2	47
Arsenic in urine	3	63
Arsenic (total) in urine	13	21
Arsenic species (As(III), As(V), monomethylarsonic acid, dimethylarsinic acid) in urine	7	97
Barium in urine	3	81
Barium, strontium, titanium in urine	2	67
Benzene	see benzene and alkylbenzenes	
Benzene	see furan-2-carboxylic acid and other carboxylic acids	
Benzene	see *t,t*-muconic acid	
Benzene	see S-phenylmercapturic acid	
Benzene and alkylbenzenes (BTX-aromatics) in blood	4	107
Benzene derivatives in urine, suitable for steam distillation (phenol; m-/p-cresol; o-cresol; o-chlorophenol; o-nitrophenol; p-chlorophenol; nitrobenzene; 1,2-dinitrobenzene; 2-chloro-5-methylphenol; 2,5-dichlorophenol; 3,4-dichlorophenol; 2,3-dinitrotoluene)	1	31
Benzidine	see haemoglobin adducts of aromatic amines	
Benzoic acid	see aromatic carboxylic acids	
Benzoic acid	see furan-2-carboxylic acid and other carboxylic acids	
Benzyl alcohol	see furan-2-carboxylic acid and other carboxylic acids	
Benzylchloride	see N-benzylvaline	
N-Benzylvaline after exposure to benzylchloride in blood	8	35

Substance	Vol.	Page
Beryllium in urine	1	57
Beryllium in urine, standard addition procedure	5	35
Beryllium, lithium, vanadium, tungsten in urine	5	51
BHT	see butylated hydroxytoluene	
Bismuth	see ICP-MS collective method	
Bisphenol A in urine	11	55
Bisphenol A, genistein, daidzein and equol in urine	13	37
Bromide in urine	1	67
Bromide in plasma and serum	10	97
2-Bromo-2-chloro-1,1,1-trifluoroethane	see halogenated hydrocarbons	
1,3-Butadiene	see monohydroxybutenylmercapturic acid (MHBMA) and dihydroxybutylmercapturic acid (DHBMA)	
1-Butanol	see alcohols and ketones	
2-Butanol	see alcohols and ketones	
2-Butanone	see alcohols and ketones	
Butoxyacetic acid in urine	4	131
2-Butoxyacetic acid	see alkoxycarboxylic acids	
2-Butoxyethanol	see alkoxycarboxylic acids	
2-Butoxyethyl acetate	see alkoxycarboxylic acids	
1-Butoxypropanol-2	see propylene and diethylene glycol ethers	
Butylated hydroxytoluene (BHT) in urine	12	105
Butyldiglycol	see 2-(2-butoxyethoxy)ethanol	
Butyltin compounds in urine (mono-, di-, tri- and tetrabutyltin)	12	121
2-Butoxyethoxyacetic acid	see alkoxycarboxylic acids	
2-(2-Butoxyethoxy)ethanol	see alkoxycarboxylic acids	
Cadmium in blood	1	79
Cadmium in urine	2	85

Substance	Vol.	Page
Cadmium	see ICP-MS collective method	
N-(2-Carbamoylethyl) mercapturic acid	see mercapturic acids	
N-(2-Carbamoylethyl)valine in blood	12	145
Carbon disulphide	see furan-2-carboxylic acid and other carboxylic acids	
Carboxyhemoglobin in blood	1	93
ChE	see acetylcholinesterase and cholinesterase	
Chlorinated aromatic hydrocarbons in plasma (o-dichlorobenzene; m-dichlorobenzene; p-dichlorobenzene; 1,2,4-trichlorobenzene; 1,2,4,5-tetrachlorobenzene; pentachlorobenzene; hexachlorobenzene)	3	93
4-Chloroaniline (p-Chloroaniline)	see haemoglobin adducts of aromatic amines	
N-(3-Chloro-2-hydroxypropyl)-valine in blood as haemoglobin adduct of epichlorohydrin	13	63
Chlorobenzenes in blood (1,2- and 1,4-dichlorobenzene; 1,2,4-trichlorobenzene)	1	107
2-Chloro-5-methylphenol	see benzene derivatives	
4-Chloro-2-methylphenoxyacetic acid	see chlorophenoxycarboxylic acids	
4-Chloro-2-methylphenoxypropionic acid	see chlorophenoxycarboxylic acids	
o-Chlorophenol, p-chlorophenol	see benzene derivatives	
Chlorophenols in urine (2,4-dichlorophenol, 2,5-dichlorophenol, 2,6-dichlorophenol, 2,3,4-trichlorophenol, 2,4,5-trichlorophenol, 2,4,6-trichlorophenol, 2,3,4,6-tetrachlorophenol)	7	143
Chlorophenols (monohydroxychlorobenzenes) in urine (2,6-, 2,3-, 3,4-dichlorophenol; 2,4,6-, 2,4,5-, 3,4,5-trichlorophenol; 2,3,4,6-, 2,3,4,5-tetrachlorophenol; pentachlorophenol)	1	123
Chlorophenoxycarboxylic acids (4-chloro-2-methylphenoxyacetic acid; 2,4-dichlorophenoxyacetic acid; 4-chloro-2-methylphenoxypropionic acid; 2,4-dichlorophenoxypropionic acid) in urine	5	77
Chlorpyrifos	see organophosphates	
Chlorpyrifos	see 3,5,6-trichloro-2-pyridinol	

Substance	Vol.	Page
Chlorpyrifos-methyl	see 3,5,6-trichloro-2-pyridinol	
4-Chloro-o-toluidine	see aromatic amines	
Cholinesterase	see acetylcholinesterase and cholinesterase	
CHPV	see N-(3-chloro-2-hydroxypropyl)-valine	
Chromium	see aluminium, chromium, cobalt, copper, manganese, molybdenum, nickel, vanadium	
Chromium in urine	2	97
Chromium in whole blood, plasma and erythrocytes	3	109
Cobalt	see aluminium, chromium, cobalt, copper, manganese, molybdenum, nickel, vanadium	
Cobalt in urine	1	141
Cobalt in blood	2	117
Copper	see aluminium, chromium, cobalt, copper, manganese, molybdenum, nickel, vanadium	
Cotinine in urine, plasma or serum	7	171
Cotinine in urine	8	53
Creatinine in urine	12	169
o-Cresol; m-/p-cresol	see benzene derivatives	
Cumene	see 2-phenyl-2-propanol	
Cyanide	see thiocyanate	
Cyanide in blood	2	133
N-2-Cyanoethylvaline, N-2-hydroxyethylvaline, N-methylvaline in blood	5	211
Cyanogen chloride	see thiocyanate	
Cyclohexane	see cyclohexanol, 1,2-cyclohexanediol and 1,4-cyclohexanediol	
Cyclohexanediol	see cyclohexanol, 1,2-cyclohexanediol and 1,4-cyclohexanediol	

Substance	Vol.	Page
Cyclohexanol, 1,2-cyclohexanediol and 1,4-cyclohexanediol in urine	13	85
Cyclohexanone	see cyclohexanol, 1,2-cyclohexanediol and 1,4-cyclohexanediol	
Cyclophosphamide	see oxazaphosphorines	
Cyfluthrin	see pyrethrum and pyrethroid metabolites	
Cypermethrin	see pyrethrum and pyrethroid metabolites	
Cytochrome P450 1A1 (genotyping)	9	67
Cytochrome P450 1A1	see short guidelines for real time PCR	
Cytochrome P450 1B1 (genotyping)	9	89
Cytochrome P450 2E1 (genotyping)	9	111
Cytochrome P450 2E1	see short guidelines for real time PCR	
Daidzein	see bisphenol A and isoflavones	
Daunorubicin	see anthracycline cytostatic agents	
DDE	see organochlorine compounds	
DDT	see organochlorine compounds	
DEHP	see di(2-ethylhexyl) phthalate (DEHP) metabolites	
Deltamethrin	see pyrethrum and pyrethroid metabolites	
DHBMA	see monohydroxybutenylmercapturic acid (MHBMA) and dihydroxybutylmercapturic acid (DHBMA)	
DHPV	see N-(2,3-dihydroxypropyl)-valine	
Dialkyl phosphates in urine (dimethyl phosphate, dimethyl thiophosphate, dimethyl dithiophosphate, diethyl phosphate, diethyl thiophosphate, diethyl dithiophosphate)	12	185
2,4-Diamino-6-chloro-s-triazine in urine	6	111

Substance	Vol.	Page
4,4′-Diaminodiphenylmethane	see aromatic amines	
4,4′-Diaminodiphenylmethane	see haemoglobin adducts of aromatic amines	
Diazinon	see organophosphates	
Dibenzodioxins	see dioxins, furans and WHO PCB	
Dibenzofurans	see dioxins, furans and WHO PCB	
cis-3-(2,2-Dibromovinyl)-2,2-dimethylcyclopropane-1-carboxylic acid	see pyrethroid metabolites	
Dibutyltin	see butyltin compounds	
1,2-Dichlorobenzene	see 3,4-dichlorocatechol and 4,5-dichlorocatechol	
1,2-Dichlorobenzene, 1,4-dichlorobenzene	see chlorobenzenes	
o-Dichlorobenzene, m-dichlorobenzene, p-dichlorobenzene	see chlorinated aromatic hydrocarbons	
3,3′-Dichlorobenzidine	see aromatic amines	
3,3′-Dichlorobenzidine	see haemoglobin adducts of aromatic amines	
3,5-Dichloroaniline	see vinclozolin	
3,4-Dichlorocatechol and 4,5-dichlorocatechol in urine	10	113
4,4′-Dichlorodiphenyldichloroethane	see organochlorine compounds	
4,4′-Dichlorodiphenyltrichloroethane	see organochlorine compounds	
1,2-Dichloroethylene	see chlorinated aromatic hydrocarbons	
Dichloromethane	see chlorinated aromatic hydrocarbons	
2,5-Dichlorophenol, 3,4-dichlorophenol	see benzene derivatives	
Dichlorophenols	see chlorophenols	
2,4-Dichlorophenoxyacetic acid	see chlorophenoxycarboxylic acids	
2,4-Dichlorophenoxypropionic acid	see chlorophenoxycarboxylic acids	

Substance	Vol.	Page
2,4-Dichlorotoluene...	see furan-2-carboxylic acid and other carboxylic acids	
cis-3-(2,2-Dichlorovinyl)-2,2-dimethylcyclopropane-1-carboxylic acid...	see pyrethroid metabolites	
trans-3-(2,2-Dichlorovinyl)-2,2-dimethylcyclopropane-1-carboxylic acid...	see pyrethroid metabolites	
Diethylene glycol dibutyl ether, diethylene glycol diethyl ether, diethylene glycol dimethyl ether, diethylene glycol monobutyl ether, diethylene glycol monobutyl ether actetate, diethylene glycol monoethyl ether, diethylene glycol monoethyl ether acetate, diethylene glycol monomethyl ether......	see propylene and diethylene glycol ethers	
Diethyl dithiophosphate...................................	see dialkyl phosphates	
Di(2-ethylhexyl) phthalate (DEHP) metabolites (2-ethyl-5-hydroxyhexyl phthalate, 2-ethyl-5-oxohexyl phthalate, mono(2-ethylhexyl) phthalate) in urine.....................	11	73
Di(2-ethylhexyl) phthalate (DEHP) metabolites in urine (mono(2-ethyl-5-hydroxyhexyl) phthalate (5OH-MEHP), mono(2-ethyl-5-oxohexyl) phthalate (5oxo-MEHP), mono(2-ethyl-5-carboxypentyl) phthalate (5cx-MEPP), mono(2-ethylhexyl) phthalate (MEHP))...............................	12	211
Diethyl phosphate...	see dialkyl phosphates	
Diethyl thiophosphate.....................................	see dialkyl phosphates	
Digestion procedures for the determination of metals in biological material...	2	1
Digestion procedures for the determination of metals in biological materials...	8	1
Dihydroxybutylmercapturic acid...........................	see monohydroxybutenylmercapturic acid (MHBMA) and dihydroxybutylmercapturic acid (DHBMA)	
3,4-Dihydroxychlorobenzene (4-chlorocatechol) in urine......	6	125
N-(2,3-Dihydroxypropyl)-valine in blood as haemoglobin adduct of glycidol...	13	101
N,N-Dimethylacetamide (DMA) and N-methylacetamide (NMA) in urine...	8	67
Dimethylarsinic acid...	see arsenic species	

Substance	Vol.	Page
Dimethylbenzoic acids in urine	12	237
N,N-Dimethylformamide (DMF) in urine	5	97
N,N-Dimethylformamide	see mercapturic acids	
N,N-Dimethylformamide	see AMCC, HMMF and NMF	
Dimethyl dithiophosphate	see dialkyl phosphates	
2,4-Dimethylnitrobenzene	see furan-2-carboxylic acid and other carboxylic acids	
2,4-Dimethylphenol, 2,3-dimethylphenol, 3,4-dimethylphenol	see phenols and aromatic alcohols	
Dimethyl phosphate	see dialkyl phosphates	
Dimethyl thiophosphate	see dialkyl phosphates	
1,2-Dinitrobenzene	see benzene derivatives	
o-Dinitrobenzene	see nitroaromatic compounds	
2,3-Dinitrotoluene	see benzene derivatives	
2,6-Dinitrotoluene	see nitroaromatic compounds	
Dioxins, furans and WHO PCB in whole blood	8	85
Dipropylene glycol monomethyl ether	see propylene and diethylene glycol ethers	
Doxorubicin	see anthracycline cytostatic agents	
Epichlorohydrine	see N-(3-chloro-2-hydroxypropyl)-valine	
Epichlorohydrine	see N-(2,3-dihydroxypropyl)-valine	
Epirubicin	see anthracycline cytostatic agents	
1,2-Epoxypropane	see mercapturic acids	
Equol	see bisphenol A and isoflavones	
Erythrocyte porphyrins (free) in blood (erythrocytes)	2	145
Ethanol	see alcohols and ketones	
2-Ethoxyacetic acid	see alkoxycarboxylic acids	
2-Ethoxyethanol	see alkoxycarboxylic acids	

Substance	Vol.	Page
2-Ethoxyethyl acetate	see alkoxycarboxylic acids	
1-Ethoxypropanol-2, 3-ethoxypropanol-1,1-ethoxypropyl acetate-2	see propylene and diethylene glycol ethers	
Ethylbenzene	see benzene and alkyl benzenes	
Ethylbenzene	see furan-2-carboxylic acid and other carboxylic acids	
Ethyl bromide	see bromide	
Ethylene oxide	see N-2-cyanoethylvaline, N-2-hydroxyethylvaline, N-methylvaline	
Ethylene oxide	see mercapturic acids	
2-Ethyl-5-hydroxyhexyl phthalate	see di(2-ethylhexyl) phthalate (DEHP) metabolites	
2-Ethyl-5-oxohexyl phthalate	see di(2-ethylhexyl) phthalate (DEHP) metabolites	
2-Ethylphenol	see phenols and aromatic alcohols	
Evaluation of susceptibility parameters in occupation and environmental medicine	9	315
Fenitrothion	see organophosphates	
Fenthion	see organophosphates	
Fluoride in urine	2	159
4-Fluoro-3-phenoxybenzoic acid	see pyrethroid metabolites	
Free erythrocyte porphyrins	see erythrocyte porphyrins (free)	
Furan-2-carboxylic acid and other carboxylic acids (phenylglyoxylic acid, mandelic acid, t,t-muconic acid, benzoic acid, hydroxybenzoic acid, hippuric acid, methylhippuric acid, 2,4-dichlorobenzoic acid, 3-methyl-4-nitro-benzoic acid, TTCA)	10	129
Furans	see dioxins, furans and WHO PCB	
Furfural	see 2-furylmethanal	
2-Furylmethanal	see furan-2-carboxylic acid and other carboxylic acids	
Gas chromatographic methods for the determination of organic substances in biological material	3	1

Substance	Vol.	Page
Glucose-6-phosphate dehydrogenase (G-6-PDH) (genotyping)	9	265
Glutathion S-transferase M1	see glutathion S-transferase T1 and M1 see short guidelines for real time PCR	
Glutathion S-transferase P1 (GSTP1) (genotyping)	9	221
Glutathion S-transferase P1	see short guidelines for real time PCR	
Glutathion S-transferase T1 (phenotyping)	9	211
Glutathion S-transferase T1	see short guidelines for real time PCR	
Glutathion S-transferase T1 and M1 (GSTT1, GSTM1) (genotyping)	9	183
Glycidol	see N-(2,3-dihydroxypropyl)-valine	
Glycol ethers, glycol ether acetates	see propylene and diethylene glycol ethers	
Gold	see platinum and gold	
Haemoglobin adducts of aromatic amines (aniline, *o*-, *m*- and *p*-toluidine, *o*-anisidine, *p*-chloroaniline, α- and β-naphthylamine, 4-aminodiphenyl, benzidine, 4,4′-diaminodiphenylmethane, 3,3′-dichlorobenzidine)	7	191
Halogenated hydrocarbons in blood (dichloromethane; 1,2-dichloroethylene; 2-bromo-2-chloro-1,1,1-trifluoroethane (halothane); trichloromethane; 1,1,1-trichloroethane; tetrachloromethane; trichloroethylene; tetrachloroethylene)	3	127
Halothane	see halogenated hydrocarbons	
HEMA	see mercapturic acids	
Hemoglobin adducts	see N-(2-carbamoylethyl)valine	
Hexachlorobenzene	see chlorinated aromatic hydrocarbons and oxachlorine compounds	
Hexachlorocyclohexane	see organochlorine compounds	
Hexahydrophthalic acid (HHP)	see hydrophthalic acids	
Hexamethylenediamine (HDA) in urine	8	119
2,5-Hexanedione	see hexane metabolites	

Substance	Vol.	Page
Hexane metabolites (2,5-hexanedione, 2-hexanone) in urine	4	147
2-Hexanone	see hexane metabolites	
2-Hexanone	see alcohols and ketones	
HHCB	see polycyclic musk compounds	
Hippuric acid	see aromatic carboxylic acids and o-, m-/p-methylhippuric acids	
Hippuric acid	see furan-2-carboxylic acid and other carboxylic acids	
HMMF	see AMCC, HMMF and NMF	
2-HMSI	see 5-hydroxy-N-methyl-2-pyrrolidone (5-HNMP) and 2-hydroxy-N-methylsuccinimide (2-HMSI)	
5-HNMP	see 5-hydroxy-N-methyl-2-pyrrolidone (5-HNMP) and 2-hydroxy-N-methylsuccinimide (2-HMSI)	
2-HPMA	see mercapturic acids	
3-HPMA	see mercapturic acids	
HPMA	see S-(3-hydroxypropyl)mercapturic acid	
Hydrazine in blood (plasma)	2	171
Hydrazine and N-acetylhydrazine in urine and plasma	6	141
Hydrogen cyanide	see thiocyanate	
Hydrophthalic acids in urine	12	261
Hydroxybenzoic acid	see furan-2-carboxylic acid and other carboxylic acids	
8-Hydroxy-2′-deoxyguanosine in urine	8	133
Hydroxyethyl mercapturic acid	see mercapturic acids	
1-(4-(1-Hydroxy-1-methylethyl)-phenyl)-3-methylurea (HMEPMU) in urine	8	151
5-Hydroxy-N-methyl-2-pyrrolidone (5-HNMP) and **2-hydroxy-N-methylsuccinimide (2-HMSI)** in urine as metabolites of N-methyl-2-pyrrolidone (NMP)	11	97

Substance	Vol.	Page
1-Hydroxyphenanthrene	see PAH metabolites	
4-Hydroxyphenanthrene	see PAH metabolites	
9-Hydroxyphenanthrene	see PAH metabolites	
2-Hydroxypropyl mercapturic acid	see mercapturic acids	
3-Hydroxypropyl mercapturic acid	see mercapturic acids	
S-(3-Hydroxypropyl)mercapturic acid (HPMA) in urine	12	281
1-Hydroxypyrene in urine	3	151
1-Hydroxypyrene	see PAH metabolites	
ICP-MS collective method (antimony, bismuth, cadmium, lead, mercury, platinum, tellurium, thallium, tin, tungsten) in urine	6	79
Idarubicin	see anthracycline cytostatic agents	
Ifosfamide	see oxazaphosphorines	
Indium in urine	3	171
Iridium in urine (Addendum to "platinum and gold")	11	115
Isoflavones	see bisphenol A and isoflavones	
Isopropylbenzene	see 2-phenyl-2-propanol	
Isoproturon	see 1-(4-(1-hydroxy-1-methyl-ethyl)-phenyl)-3-methylurea (HMEPMU)	
Lead in blood	1	155
Lead in blood and urine	2	183
Lead	see ICP-MS collective method	
Lithium	see beryllium, lithium, vanadium, tungsten	
Malathion	see organophosphates	
Mandelic acid	see aromatic carboxylic acids	
Mandelic acid	see furan-2-carboxylic acid and other carboxylic acids	
Manganese	see aluminium, chromium, cobalt, copper, manganese, molybdenum, nickel, vanadium	

Substance	Vol.	Page
Manganese in blood	10	157
MEHP	see di(2-ethylhexyl) phthalate (DEHP) metabolites	
Mercapturic acids (N-acetyl-S-(2-carbamoylethyl)-L-cysteine, N-acetyl-S-(2-hydroxyethyl)-L-cysteine, N-acetyl-S-(3-hydroxypropyl)-L-cysteine, N-acetyl-S-(2-hydroxypropyl)-L-cysteine, N-acetyl-S-(N-methylcarbamoyl)-L-cysteine) in urine	13	123
Mercury in blood and urine	2	195
Mercury	see ICP-MS collective method	
Metasystox R in urine	7	221
Methanol	see alcohols and ketones	
Methoxyacetic acid	see alkoxycarboxylic acids	
Methoxyethanol	see alkoxycarboxylic acids	
Methoxyethyl acetate	see alkoxycarboxylic acids	
1-Methoxypropanol-2	see propylene and diethylene glycol ethers	
2-Methoxypropanol	see alkoxycarboxylic acids	
1-Methoxypropanone-2	see propylene and diethylene glycol ethers	
2-Methoxypropionic acid	see alkoxycarboxylic acids	
1-Methoxypropyl acetate-2	see propylene and diethylene glycol ethers	
2-Methoxypropyl acetate-1	see alkoxycarboxylic acids	
N-Methylacetamide	see N,N-dimethylacetamide (DMA) and N-methylacetamide (NMA)	
Methylating agents (bis(chloromethyl)ether, bromomethane, chloromethane, dimethyl sulphate, iodomethane, monochlorodimethyl ether)	see N-2-cyanoethylvaline, N-2-hydroxyethylvaline, N-2-methylvaline	
3-Methylbenzyl alcohol	see phenols and aromatic alcohols	
Methyl bromide	see bromide	
4,4′-Methylene-bis(2-chloroaniline)	see aromatic amines	
4-Methylhexahydrophthalic acid (4-MHHP)	see hydrophthalic acids	

Substance	Vol.	Page
Methylhippuric acid	see furan-2-carboxylic acid and other carboxylic acids	
o-Methylhippuric acid, *m-/p*-methylhippuric acids (toluric acids), hippuric acid in urine	1	165
o-Methylhippuric acid, *m-/p*-methylhippuric acids	see aromatic carboxylic acids	
Methylmercury in blood	10	169
3-Methyl-4-nitro-benzoic acid	see furan-2-carboxylic acid and other carboxylic acids	
4-Methyl-2-pentanone	see alcohols and ketones	
2-Methylphenol, 4-methylphenol	see phenols and aromatic alcohols	
2-Methyl-1-propanol	see alcohols and ketones	
N-Methyl-2-pyrrolidone (NMP)	see 5-hydroxy-*N*-methyl-2-pyrrolidone (5-HNMP) and 2-hydroxy-*N*-methylsuccinimide (2-HMSI)	
3-Methyltetrahydrophthalic acid (3-MTHP)	see hydrophthalic acids	
4-Methyltetrahydrophthalic acid (4-MTHP)	see hydrophthalic acids	
MHBMA	see monohydroxybutenylmercapturic acid (MHBMA) and dihydroxybutylmercapturic acid (DHBMA)	
β_2-**Microglobulin** in urine and serum	3	185
Molybdenum	see aluminium, chromium, cobalt, copper, manganese, molybdenum, nickel, vanadium	
Molybdenum in urine	5	109
Molybdenum in plasma and urine	8	167
Monobutyltin	see butyltin compounds	
Mono(2-ethylhexyl) phthalate	see di(2-ethylhexyl) phthalate (DEHP) metabolites	
Monohydroxybutenylmercapturic acid (MHBMA) and dihydroxybutylmercapturic acid (DHBMA) in urine as metabolites of 1,3-butadiene	11	127
Monohydroxychlorobenzenes	see chlorophenols	

Substance	Vol.	Page
Monomethylarsonic acid	see arsenic species	
t,t-Muconic acid	see furan-2-carboxylic acid and other carboxylic acids	
t,t-Muconic acid in urine	5	125
Musk compounds	see polycyclic musk compounds	
Naphthalene metabolites 1-naphthol and 2-naphthol in urine	12	301
1-Naphthol	see naphthalene metabolites	
2-Naphthol	see naphthalene metabolites	
1-Naphthylamine, 2-naphthylamine	see aromatic amines	
1-Naphthylamine, 2-naphthylamine	see haemoglobin adducts of aromatic amines	
α-Naphthylamine, β-naphthylamine	see haemoglobin adducts of aromatic amines	
Neuropathy target esterase (NTE) in leukocytes	11	151
Nickel	see aluminium, chromium, cobalt, copper, manganese, molybdenum, nickel, vanadium	
Nickel in blood	3	193
Nickel in urine	1	177
Nitroaromatic compounds in plasma (nitrobenzene; *p*-nitrotoluene, *p*-nitrochlorobenzene; 2,6-dinitrotoluene; *o*-dinitrobenzene; 1-nitronaphthalene; 2-nitronaphthalene; 4-nitrobiphenyl)	3	207
Nitrobenzene	see benzene derivatives	
Nitrobenzene	see nitroaromatic compounds	
4-Nitrobiphenyl	see nitroaromatic compounds	
p-Nitrochlorobenzene	see nitroaromatic compounds	
1-Nitronaphthaline, 2-nitronaphthaline	see nitroaromatic compounds	
o-Nitrophenol	see benzene derivatives	
p-Nitrotoluene	see nitroaromatic compounds	

Substance	Vol.	Page
NMP	see 5-hydroxy-N-methyl-2-pyrrolidone (5-HNMP) and 2-hydroxy-N-methylsuccinimide (2-HMSI)	
NTE	see neuropathy target esterase	
Ochratoxin A in plasma und serum	12	341
Organochlorine compounds in whole blood and plasma	8	187
Organophosphates in whole blood (chlorpyrifos, diazinon, fenitrothion, fenthion, malathion)	11	167
Organotin compounds (except methyltin compounds) and total tin in urine	4	165
Oxalic acid dinitrile	see thiocyanate	
Oxazaphosphorines (cyclophosphamide and ifosfamide) in urine	8	221
Oxydemeton-methyl in urine	see metasystox R	
PAH metabolites in urine	6	163
Palladium in urine	11	189
Pentachlorobenzene	see chlorinated aromatic hydrocarbons	
Pentachlorophenol	see chlorophenols	
Pentachlorophenol in urine	7	237
Pentachlorophenol in urine and serum/plasma	6	189
Perfluorooctanoic acid in plasma	10	191
Perfluorooctanesulphonic acid and Perfluorobutanesulphonic acid in plasma and urine	10	213
Permethrin	see pyrethrum and pyrethroid metabolites	
PFBS	see perfluorooctanesulphonic acid and perfluorobutanesulphonic acid	
PFOA	see perfluorooctanoic acid	
PFOS	see perfluorooctanesulphonic acid and perfluorobutanesulphonic acid	

Substance	Vol.	Page
Phenol in urine	1	189
Phenol	see benzene derivatives and phenols and aromatic alcohols	
Phenols and aromatic alcohols in urine (phenol; 2- and 4-methylphenol; DL-1- and 2-phenylethanol; 3-methylbenzyl alcohol; 2-ethylphenol; 2,4-; 2,3- and 3,4-dimethylphenol)	2	213
Phenols in urine	6	211
Phenothrin	see pyrethrum and pyrethoid metabolites	
Phenoxyacetic acid	see alkoxycarboxylic acids	
3-Phenoxybenzoic acid	see pyrethroid metabolites	
2-Phenoxyethanol	see alkoxycarboxylic acids	
1-Phenylethanol, 2-phenylethanol	see phenols and aromatic alcohols	
Phenylglyoxylic acid	see aromatic carboxylic acids	
Phenylglyoxylic acid	see furan-2-carboxylic acid and other carboxylic acids	
2-Phenyl-2-propanol in urine	13	163
S-Phenylmercapturic acid in urine	5	143
Platinum in urine, blood, plasma/serum	4	187
Platinum	see ICP-MS collective method	
Platinum and gold in urine	7	255
Polychlorinated biphenyls in blood or serum	3	231
Polychlorinated biphenyls	see dioxins, furans and WHO PCB	
Polychlorinated biphenyls	see organochlorine compounds	
Polychlorinated dibenzodioxines	see dioxins, furans and WHO PCB	
Polychlorinated dibenzofurans	see dioxins, furans and WHO PCB	

Substance	Vol.	Page
Polycyclic musk compounds (PMC) in blood (1,3,4,6,7,8-hexahydro-4,6,6,7,8,8-hexamethylcyclopenta[g]-2-benzopyrane (HHCB), 7-acetyl-1,1,3,4,4,6-hexamethyltetrahydro naphthalene (AHTN), 4-acetyl-1,1-dimethyl-6-tert-butylindane (ADBI) and 6-acetyl-1,1,2,3,3,5-hexamethylindane (AHDI)) . . .	11	209
Polymerase chain reaction and its application in occupational and environmental medicine	9	35
Potassium cyanide	see thiocyanate	
1-Propanol	see alcohols and ketones	
2-Propanol	see alcohols and ketones	
Propenal	see mercapturic acids	
Propylene and diethylene glycol ethers in urine and blood (1-butoxypropanol-2, diethylene glycol dibutyl ether, diethylene glycol diethyl ether, diethylene glycol dimethyl ether, diethylene glycol monobutyl ether, diethylene glycol monobutyl ether acetate, diethylene glycol monoethyl ether, diethylene glycol monoethyl ether acetate, diethylene glycol monomethyl ether, dipropylene glycol monomethyl ether, 1-ethoxypropanol-2, 3-ethoxypropanol-1, 1-ethoxypropyl acetate-2, 1-methoxypropanol-2, 1-methoxypropanone-2, 1-methoxypropyl acetate-2, propylene glycol diacetate)	11	231
Propylene glycol diacetate	see propylene and diethylene glycol ethers	
Propylene oxide	see mercapturic acids	
Pyrethroid metabolites (*cis*-3-(2,2-dichlorovinyl)-2,2-dimethylcyclopropane-1-carboxylic acid, *trans*-3-(2,2-dichlorovinyl)-2,2-dimethylcyclopropane-1-carboxylic acid, *cis*-3-(2,2-dibromovinyl)-2,2-dimethylcyclopropane-1-carboxylic acid, 4-fluoro-3-phenoxybenzoic acid, 3-phenoxybenzoic acid) in urine	6	231
Pyrethrum and pyrethroid metabolites (after solid phase extraction) in urine	13	179
Pyrethrum and pyrethroid metabolites (after liquid-liquid extraction) in urine	13	215
Resmethrin	see pyrethrum and pyrethroid metabolites	
Rhodium in urine and serum/plasma	7	273
Selenium in blood, plasma and urine	2	231

Substance	Vol.	Page
Selenium in serum	13	249
Short guidelines for real-time PCR	9	281
Sodium cyanide	see thiocyanate	
Strontium	see barium, strontium, titanium	
Styrene	see furan-2-carboxylic acid and other carboxylic acids	
Sulphotransferase 1A1 and 1A2 (genotyping)	9	241
Susceptibility and biological monitoring	9	1
Susceptibility and determination methods	9	5
TCDD	see 2,3,7,8-tetrachlorodibenzo-p-dioxin	
TCPyr	see 3,5,6-trichloro-2-pyridinol	
Tellurium	see ICP-MS collective method	
Tetrabutyltin	see butyltin compounds	
1,2,4,5-Tetrachlorobenzene	see chlorinated aromatic hydrocarbons	
2,3,7,8-Tetrachlorodibenzo-p-dioxin in blood	11	261
Tetrachloroethylene	see halogenated hydrocarbons	
Tetrachloromethane	see halogenated hydrocarbons	
2,3,4,6-Tetrachlorophenol, 2,3,4,5-tetrachlorophenol	see chlorophenols	
Tetrahydrophthalic acid (THP)	see hydrophthalic acids	
Tetrahydrofuran (THF) in urine (Addendum to the DFG method "Alcohols and Ketones")	13	265
Tetramethrin	see pyrethrum and pyrethoid metabolites	
Thallium in urine	1	199
Thallium in urine	5	163
Thallium	see ICP-MS collective method	
The use of atomic absorption spectrometry for the determination of metals in biological materials	4	1
The use of liquid chromatography/mass spectrometry (LC/MS) in biological monitoring	11	3

Substance	Vol.	Page
Thiocyanate in plasma and saliva	13	277
2-Thioxothiazolidine-4-carboxylic acid (TTCA) in urine	4	207
Thorium and uranium in urine	6	255
Tin in urine	4	223
Tin	see ICP-MS collective method	
Titanium	see barium; strontium, titanium	
Titanium in blood and urine	10	237
Toluene	see benzene and alkylbenzenes	
Toluene	see furan-2-carboxylic acid and other carboxylic acids	
o-/m-/p-Toluidines	see haemoglobin adducts of aromatic amines	
o-Toluidine, m-toluidine, p-toluidine	see aromatic amines	
Toluric acids	see methylhippuric acids	
2,4- and 2,6-Toluylendiamine	see aromatic amines	
Tributyltin	see butyltin compounds	
Trichloroacetic acid (TCA) in urine	1	209
1,2,4-Trichlorobenzene	see chlorobenzenes and chlorinated aromatic hydrocarbons	
Trichloroethane	see halogenated hydrocarbons	
Trichloroethylene	see halogenated hydrocarbons	
Trichloromethane	see halogenated hydrocarbons	
Trichlorophenols	see chlorophenols	
3,5,6-Trichloro-2-pyridinol in urine	10	253
Trimethylbenzene	see dimethylbenzoic acids	
Trinitrotoluene	see aminodinitrotoluenes	
Tungsten	see beryllium, lithium, vanadium, tungsten	
Tungsten	see ICP-MS collective method	
Uranium	see thorium and uranium	

Substance	Vol.	Page
Vanadium	see aluminium, chromium, cobalt, copper, manganese, molybdenum, nickel, vanadium	
Vanadium in urine	3	241
Vanadium	see beryllium, lithium, vanadium, tungsten	
Vinclozolin as 3,5-dichloroaniline in urine	7	287
Xylene	see furan-2-carboxylic acid and other carboxylic acids	
o-Xylene, m-xylene	see benzene and alkylbenzenes	
Zinc in plasma, serum and urine	5	211

The working group "Analyses in Biological Materials" of the DFG Senate Commission for the Investigation of Health Hazards of Chemical Compounds in the Work Area

Organisation

The Senate Commission for the Investigation of Health Hazards of Chemical Compounds in the Work Area was founded in 1955 by the *Deutsche Forschungsgemeinschaft* (DFG, German Research Foundation). The DFG Senate Commission decided to constitute the working group "Analyses in Biological Materials" in 1972. The working group consists of experienced scientists from national and international universities, research institutions and industry. The participants are nominated by the DFG Senate Commission as permanent members or guests for three years each. Furthermore, ad-hoc experts can be invited by the working group's leader to make contributions on current topics. All participants of the working group are active on a voluntary basis.

At present, the working group is chaired by Prof. Dr. Thomas Göen. A list of the current members, guests and ad-hoc experts of the working group "Analyses in Biological Materials" is given at the end of this volume.

The working group's tasks

The aim of the working group is to promote the development and publication of analytical methods for the determination of hazardous chemical substances, their metabolites or exposure parameters in biological materials. The main objective here is the quantitation of substances in concentration ranges relevant in occupational medicine for monitoring occupational exposure. To an increasing extent, however, methods are being developed for use in environmental medicine, too. In addition, the latter can be used for establishing and monitoring reference values and the background exposure of the general population.

In addition to developing analytical methods, the working group also deals with general analytical questions. This involves both the presentation of essential procedures for instrumental analytics as well as summaries on requirements in the pre-analytical period and on the statistical evaluation of analytical results. The general chapters are published in the Preliminary Remarks of the "MAK Collection. Part IV: Biomonitoring Methods".

The work flow within the working group

The selection of biomarkers and of methods that are to be developed and examined is carried out, among others, in close cooperation with the DFG working groups

"Derivation of MAK Values" (MAK group) and "Settingt of Threshold Values in Biological Material" (BAT group). The development of suitable analytical methods is based on relevant scientific literature and the expertise and competency of the members and guests of the working group. In general practice, a member of the working group submits a method to the working group, which has been developed in his/her laboratory. If no suitable analytical procedures for special parameters are available, a member of the group is asked to develop such a method. Following a discussion in the working group, at least one examiner is appointed to reproduce the analytical method in his/her laboratory. As a matter of principle, the method examination includes all described analytical procedures, whereby special attention is paid to the examination of the given validation data. The examiner creates a detailed examination report, which is then discussed in the group. As soon as a positive consensus is reached by the working group, the method is adopted and passed on for publication.

Generally, only those methods are adopted and published, which meet the analytical reliability criteria (such as accuracy, precision, sensitivity and specificity) confirming the quality of the method and the special requirements of analytical procedures for environmental and occupational medicine. The standard working procedure of the working group (i.e. development of the methods by qualified and competent analytical experts and subsequent reproduction and examination of the methods) guarantees that only reliable analytical procedures are published, which can be reproduced with the given reliability criteria. As a matter of principle, a reliable identification and quantification of the hazardous substances or its metabolites has priority over other aspects such as easiness or economy.

Publications by the working group

The adopted methods are published as standard operating procedures (SOP). The SOPs list the equipment and chemicals to be used, gives the preparation and storage of the solutions, a detailed description of specimen collection and sample preparation as well as information on the calibration and the calculation of the analytical results. Furthermore, details on reliability criteria as well as a comprehensive discussion of the method are included. At the beginning of every method there is a general chapter. Therein, the toxicology of the hazardous substance or its metabolite(s) is discussed. Moreover, available data on the respective biomonitoring parameters are presented. In general, different methods may be published for one and the same substance or its metabolite(s). This is meant, among others, to account for the fact that the equipment or instruments used in practice may vary according to the possibilities given.

The adopted methods are published by the DFG at regular intervals in German in a loose-leaf collection entitled *"Analytische Methoden zur Prüfung gesundheitsschädlicher Arbeitsstoffe"* (Wiley VCH, Weinheim). The collection is arranged according to subject areas and consists of two volumes:

- Volume 1: "Analyses in Air"
- Volume 2: "Analyses in Biological Materials"

Additionally, the methods are also published in English in the book series "The MAK Collection. Part IV: Biomonitoring Methods" (Wiley VCH, Weinheim).

Since January 2012 all publications of the DFG Commission are also available in open access: http://onlinelibrary.wiley.com/book/10.1002/3527600418/topics

Withdrawal of methods

Already published analytical procedures which have lost their validity due to new developments in the instrumental analytics or that are out of date based on new knowledge in occupational medical toxicology can be replaced by more efficient methods. After consultation of the working group such methods are withdrawn by the chair of the working group and marked as such in the German loose-leaf collection.

Terms and symbols used

Terminology

Accuracy – the numerical difference between the mean of a set of replicate measurements and the true value.
Analyte – the substance to be measured.

Biological Tolerance Value (BAT, *Biologischer Arbeitsstoff-Toleranzwert*) – is defined as the occupational-medical and toxicological derived concentration for a substance, its metabolites or an effect parameter in the corresponding biological material at which the health of an employee generally is not adversely affected even when the person is repeatedly exposed during long periods. BAT values are based on a relationship between external and systemic exposure or between the systemic exposure and the resulting effect of the substance. The derivation of the BAT value is based on the average of systemic exposures.
Blank value – the analytical result obtained when the complete procedure is carried out on ultrapure water instead of biological specimens.

Calibration standards – specimens with known concentrations of the analyte that are used for calibration.
Certified value – the concentration of the analyte in control specimens certified by an official body using conditions established by that body.
Control material – a material that is used solely for quality control purposes and not for calibration.

Detection limit – the minimum analytical result that is still clearly detectable and distinguishable from the background noise.

External quality control – analyses performed on the same specimen by several laboratories for quality control purposes.

Interference – the effect on the accuracy of measurement of one analyte caused by another substance that does not itself produce a signal.
Internal quality control – the determination of precision and accuracy of a method by one laboratory for quality control purposes.

Maximum Concentration Value at the Workplace (MAK, *Maximale Arbeitsplatzkonzentration*) – is defined as the maximum concentration of a chemical substance (as gas, vapour or particulate matter) in the workplace air which generally does not have known adverse effects on the health of the employee nor cause unreasonable annoyance (e.g. by a nauseous odour) even when the person is repeatedly exposed during long periods, usually for 8 hours daily but assuming on average a 40-hour working week.

Preanalytical phase – the period from the specimen collection to the aliquotation (sampling) of the biological specimens.
Precision – the standard deviation or coefficient of variation of the results in a set of replicate measurements. A distinction is made between within day and day to day precision.
Prognostic range – an interval that with a given probability includes the analytical result from an identical specimen.

Quality control – the assessment of accuracy and reliability of laboratory measurement processes.
Quantitation limit (Limit of Quantification, LOQ) – The lowest concentration of an analyte that can be quantitatively determined with a defined standard deviation (as a rule 33.3%) (also lower limit of quantification/quantitation, LLOQ).

Recovery rate – in recovery experiments the amount of recovered analyte divided by the amount of added analyte expressed as a percentage.
Reference material – a material or substance for which one or more properties are sufficiently well established to be used for calibration of an apparatus or for verification of an analytical method.
Reference method – an analytical method of proven and demonstrated accuracy and precision.
Reliability criteria – defined quantifiable parameters for the assessment of the quality of an analytical method, e.g., precision, accuracy, detection limit.

Sample – a representative quantity of material for measurement of the analyte concentration.
Sensitivity – relates to an analytical method's ability to identify positive results.
Specificity – The capability of a method to detect and unambiguously identify the analyte without interference from other matrix components, The term selectivity is often used synonymously.
Specimen – the material available for analysis.

Terms and symbols used

Symbols

c	substance concentration of analyte
$\bar{c}$	mean substance concentration of analyte
E	extinction
m	mass
M	molar mass
n	number
p	pressure
P	probability
r	recovery rate
s	relative standard deviation
s_w	relative standard deviation derived from replicate analyses of the same specimen
t_p	Students t factor
T	thermodynamic temperature
u	prognostic range
V	volume
V_m	molar volume
x	observed measure of analyte
$\bar{x}$	mean
z	number of duplicate analyses

Analytical Methods

N-Acetyl-S-(N-methylcarbamoyl)-cysteine (AMCC), N-hydroxymethyl-N-methyl-formamide (HMMF) and N-methyl-formamide (NMF) in urine

Matrix:	Urine
Hazardous substances:	N,N-Dimethylformamide
Analytical principle:	Capillary gas chromatography with nitrogen selective detection (GC-NPD, GC-TSD)
Completed in:	December 2005

Overview of parameters that can be determined with this method and the corresponding chemical substances:

Hazardous Substance	CAS	Parameter	CAS
N,N-Dimethyl-formamide	68-12-2	N-Hydroxymethyl-N-methyl-formamide (HMMF)	20546-32-1
		N-Acetyl-S-(N-methylcarbamoyl)-cysteine (AMCC)	103974-29-4
		N-Methylformamide (NMF)	123-39-7

Summary

The present method permits sensitive and specific quantitation of N-acetyl-S-(N-methylcarbamoyl)-cysteine (AMCC) as well as the sum of both N-hydroxy-N-methylformamide (HMMF) and N-methylformamide (NMF) as metabolites of N,N-dimethylformamide (DMF) in urine. With this method, the internal exposure of persons exposed to DMF at their workplaces can be determined.

After acidification, the urine sample is extracted twice with tetrahydrofuran, and the resulting organic extract cleaned up using a cation exchange resin. The solvent is reduced by evaporation and the residue is then dissolved in ethanol. AMCC is converted to the analyte ethyl-N-methylcarbamate (EMC) in the presence of potassium carbonate. The analytes are separated by capillary gas chromatography and subjected to nitrogen selective detection (nitrogen-phosphorus detector (NPD) or thermionic specific detector (TSD)). HMMF decomposes in the injector to N-methylformamide (NMF), which is detected. Calibration standards are prepared in urine and are treated in the same manner as the samples to be investigated. NMF is used as the calibration standard for HMMF, with N,N-dimethylpropionic acid amide (DMPA) as the internal standard.

HMMF + NMF

Within day precision:	Standard deviation (rel.)	$s_w = 2.6\%$
	Prognostic range	$u = 7.2\%$
	at a spiked concentration of 30 mg NMF per litre urine and where n = 5 determinations	
Day to day precision:	Standard deviation (rel.)	$s_w = 5.3\%$
	Prognostic range	$u = 14.7\%$
	at a spiked concentration of 30 mg NMF per litre urine and where n = 5 determinations	
Accuracy:	Recovery rate	$r = 94.6\%$
	at a nominal concentration of 25 mg NMF per litre urine and where n = 8 determinations	
Detection limit:	1.0 mg NMF per litre urine	
Quantitation limit:	3.0 mg NMF per litre urine	

AMCC

Within day precision:	Standard deviation (rel.)	$s_w = 9.4\%$
	Prognostic range	$u = 26.1\%$
	at a spiked concentration of 40 mg AMCC per litre urine and where n = 5 determinations	
Day to day precision:	Standard deviation (rel.)	$s_w = 10.8\%$
	Prognostic range	$u = 30.0\%$
	at a spiked concentration of 40 mg AMCC per litre urine and where n = 5 determinations	
Accuracy:	Recovery rate	$r = 81.4\%$
	at a nominal concentration of 25 mg AMCC per litre urine and where n = 8 determinations	
Detection limit:	0.5 mg AMCC per litre urine	
Quantitation limit:	1.5 mg AMCC per litre urine	

N,N-Dimethylformamide (DMF)

N,N-Dimethylformamide (73.1 g/mol) is a colourless polar organic solvent with a boiling point of 153°C. On account of its almost unlimited miscibility with water and other organic compounds, it is a popular solvent used in both, the pharmaceutical and chemical industries. At the workplace, DMF is mostly absorbed via the lungs and skin. However, due to the closed production processes during manufacture of DMF, exposures are limited. On the other hand, during further industrial processing (e.g. production of synthetic leather or polyacrylic fibres) considerable exposure levels may occur due to "open" processes.

In humans, DMF is metabolised to several products including N-hydroxymethyl-N-methylformamide (HMMF), N-methylformamide (NMF) and N-acetyl-S-(N-methylcarbamoyl)-cysteine (AMCC) [1]. AMCC is more commonly formed in humans than in rodents [2]. Methylisocyanate, an intermediary product, is assumed to be responsible for the toxic effects of DMF [3, 4]. Figure 1 (see appendix) gives an overview of the metabolism of DMF.

All three parameters – HMMF, NMF and AMCC – are suitable biomarkers for biological monitoring of persons exposed to DMF [5–7]. In all gas chromatographic analytical procedures, the sum of HMMF and NMF is determined in the form of NMF, as HMMF decomposes to NMF and formaldehyde in the injector port of the gas chromatograph (GC) at temperatures higher than 250°C (see Figure 2). On account of the skin penetration properties of DMF, biological monitoring is suitable for monitoring the BAT value of 35 mg NMF per litre urine [8].

Due to its toxicity [9, 10], exposure to DMF at the workplace is subject to legal regulation. In Germany, the maximum concentration at the workplace (MAK) for DMF in the air must not exceed 5 ppm (15 mg/m^3). Moreover, DMF is designated with an "H" because of possible skin absorption at the workplace. DMF is classified in pregnancy risk group B [8], as embryonic or foetal damage cannot be excluded in cases where women are exposed to DMF during pregnancy, even when the MAK value is not exceeded. This is because dermal absorption at the workplace is unavoidable – in spite of intensive protective measures. The biological tolerance value (BAT value) for the sum of HMMF and NMF (measured as NMF) of 35 mg/L at the end of a shift is thus independent of the corresponding MAK value. In the USA, the biological exposure index (BEI) value for AMCC is set at a level of 40 mg/L (measured in pre-shift urine samples at the end of a working week).

Authors: *H. U. Käfferlein, J. Angerer*
Examiner: *G. Leng*

N-Acetyl-S-(N-methylcarbamoyl)-cysteine (AMCC), N-hydroxymethyl-N-methylformamide (HMMF) and N-methylformamide (NMF) in urine

Matrix:	Urine
Hazardous substances:	N,N-Dimethylformamide
Analytical principle:	Capillary gas chromatography with nitrogen selective detection (GC-NPD, GC-TSD)
Completed in:	December 2005

Contents

1	General principles
2	Equipment, chemicals and solutions
2.1	Equipment
2.2	Chemicals
2.3	Solutions
2.4	Internal standard
2.5	Calibration standards
3	Specimen collection and sample preparation
3.1	Specimen collection
3.2	Sample preparation
4	Operational parameters
5	Analytical determination
6	Calibration
7	Calculation of the analytical result
8	Standardisation and quality control
9	Evaluation of the method
9.1	Precision
9.2	Accuracy
9.3	Detection limits
9.4	Sources of error

10	Discussion of the method
11	References
12	Appendix

1 General principles

After acidification and saturation with sodium chloride, the urine sample is extracted twice with tetrahydrofuran (THF). The organic extract is then cleaned up using a cation exchange resin. After evaporation the residue is dissolved in ethanol. AMCC is then converted to the analyte ethyl-N-methylcarbamate (EMC) in the presence of potassium carbonate (see Figure 3). Finally, the analytes are separated by capillary gas chromatography and subjected to detection using a nitrogen selective detector (NPD or TSD). At this point, N-hydroxy-N-methylformamide (HMMF) decomposes in the injector to N-methylformamide (NMF) (see Figure 2). Calibration standards are prepared in urine and are treated in the same manner as the samples to be investigated. NMF is used as the calibration standard for HMMF and N,N-dimethylpropionic acid amide (DMPA) as an internal standard.

2 Equipment, chemicals and solutions

2.1 Equipment

- Gas chromatograph with nitrogen selective detector (NPD, TSD), split-/splitless injector, autosampler and data processing system
- Capillary gas chromatographic column: length: 60 m; inner diameter: 0.25 mm; stationary phase: 15% dimethylpolysiloxane, 85% polyethylene glycol, film thickness: 0.25 µm (e.g. DX-4 of J&W Scientific, Folsom, USA)
- Thermostatic evaporation unit (e.g. by Labor-Technik Barkey)
- Laboratory shaker (e.g. by Sarstedt)
- Laboratory centrifuge (e.g. by Heraeus)
- Adjustable pipettes, 100 µL, 1 mL and 5 mL (e.g. by Eppendorf)
- 100 mL and 1000 mL Volumetric flasks
- 1.8 mL Glass vials (e.g. by Agilent)
- 20 mL Screw cap glass tubes with septa (e.g. by Schott)
- 5 mL Screw cap glass tubes with septa (e.g. by Schott)
- Polypropylene container for urine collection (e.g. by Kautex)

2.2 Chemicals

- N-Acetyl-S-(N-methylcarbamoyl)-cysteine (e.g. Sigma-Aldrich, No. 90914)
- N,N-Dimethylpropionic acid amide, 98% (e.g. Sigma-Aldrich, No. 252875)
- Ethyl-N-methylcarbamate, 98% (e.g. Tokyo Kasei, No. M0456)
- N-Methylformamide, 99% (e.g. Sigma-Aldrich, No. M46705)
- Cation exchange resin AG 50W-X8 (100-200 mesh) (Bio-Rad, No. 142-1441)
- Water, bidistilled
- Hydrochloric acid 25% (e.g. Merck, No. 100316)
- Tetrahydrofurane, "stabiliser free" (e.g. Merck, No. 108107)
- Sodium chloride p.a. (e.g. Merck, No. 106404)
- Potassium carbonate p.a. (e.g. Merck, No. 104928)
- Sodium sulphate, anhydrous (e.g. Merck, No. 106649)
- Dichloromethane p.a. (e.g. Merck, No. 1.06054)
- Ethanol absolute p.a. (e.g. Merck, No. 1.00983)
- Nitrogen 5.0
- Helium 5.0

2.3 Solutions

- 1 N Hydrochloric acid
  128 mL of the 25% hydrochloric acid are placed into a 1000 mL volumetric flask. The flask is then made up to the mark with bidist. water.

2.4 Internal standard

- Working solution of the internal standard (100 mg/L)
  10 mg dimethylpropionic acid amide (DMPA) are weighed into a 100 mL volumetric flask. The flask is then made up to the mark with bidistilled water.

2.5 Calibration standards

- Stock solution NMF (100 mg/L)
  10 mg N-methylformamide (NMF) are weighed into a 100 mL volumetric flask. The flask is then made up to the mark with pooled urine.
- Stock solution AMCC (100 mg/L)
  10 mg N-acetyl-S-(N-methylcarbamoyl)-cysteine (AMCC) are weighed into a 100 mL volumetric flask. The flask is then made up to the mark with pooled urine.

Table 1 Pipetting scheme for the preparation of calibration standards in pooled urine.

Volume of the stock solution NMF [mL]	Volume of the stock solution AMCC [mL]	Pooled urine or water [mL]	Analyte level of calibration standard (NMF or AMCC) [mg/L]
–	–	100	0
1	1	98	1
5	5	90	5
10	10	80	10
25	25	50	25
50	50	0	50

From the stock solutions of NMF and AMCC calibration standards with concentrations between 1.0 and 50.0 mg/L are prepared by diluting with pooled urine according to Table 1. For this purpose, the given quantities of each stock solution are placed into a 100 mL volumetric flask and the flasks are made up to the mark with pooled urine. From these standard solutions, aliquots of 6 mL each are pipetted into sealable 20 mL screw cap glass tubes and frozen at –20°C. Standards stored in this way can be kept for at least 6 months without noteworthy analyte losses.

As no matrix effects occur, water can be used for calibration instead of pooled urine. Furthermore, as pooled urine may contain background levels of AMCC, calibration in water can be advantageous.

3 Specimen collection and sample preparation

3.1 Specimen collection

Urine samples of persons occupationally exposed to DMF and of control persons are collected in sealable polypropylene containers and kept frozen at –20°C. Samples stored in this manner can be kept for several years.

3.2 Sample preparation

For sample preparation please note the details listed under Section 9.4 ("Sources of Error").

The urine samples are thawed and equilibrated to room temperature. In a 20 mL screw cap glass tube with septum, 2.5 g sodium chloride, 500 µL 1 N hydrochloric acid and 100 µL of the working solution of the internal standard (IS) are added to 5 mL urine sample, and the sample is shaken thoroughly for 5 min on a laboratory shaker. Then, the sample is extracted twice using 5 mL tetrahydrofurane (THF) each time. For this purpose, the sample is centrifuged at 2600 g for 5 min and the organic phase transferred to a new 20 mL screw cap glass tube with septum, in which 1 g of the cation exchange resin has already been placed. The combined organic phases are shaken using a laboratory shaker at a low mixing frequency for 60 min and then centrifuged for 10 min at 800 g. The supernatant is removed with a pipette and transferred to a new 20 mL screw cap glass tube. The cation exchange resin is subsequently washed by adding 3 mL THF and centrifuged again. The resulting supernatant is then also transferred to the 20 mL screw cap glass tube. After combining the organic phases, the 20 mL screw cap glass tube is placed in an evaporator and the THF is reduced by evaporation to a volume of about 500 µL in a stream of nitrogen under gentle heating (to compensate for cooling from evaporation). After equilibration of the sample to room temperature, 2 mL ethanol are added and the sample is briefly shaken by hand, before transferring to a 5 mL screw cap glass tube filled with 1.5 g potassium carbonate. The glass tube is immediately sealed with screw cap and septum and thoroughly mixed for one minute using a laboratory shaker. Then, all further sample solutions are separately transferred in this manner to screw cap glass tubes filled with potassium carbonate, mixed, and finally centrifuged for 5 min at 2600 g. 1 mL each is removed from the supernatant, transferred to a 1.8 mL glass vial and sealed. Finally the sample is separated by capillary gas chromatography and the metabolites HMMF and AMCC are detected in the form of NMF and EMC using nitrogen selective detection (NPD, TSD).

4 Operational parameters

Capillary column:	Material:	Fused Silica
	Stationary phase:	DX-4
	Length:	60 m
	Inner diameter:	0.25 mm
	Film thickness:	0.25 µm
Detector:	Nitrogen selective detector (NPD, TSD)	
Temperature:	Column:	10 min at 100°C; then increase at 3°C/min to 140°C; hold for 11 min, then increase at 25°C/min to 240°C; hold 15 min at final temperature
	Injector:	280°C
	Detector:	280°C
Carrier gas:	Helium 5.0 with 20 psi (constant)	
Sample volume:	1 µL (splitless, split after 60 s)	
Bead power:	3.0 V	
Make-up gas:	Nitrogen 5.0	

All other parameters must be optimised in accordance with the manufacturer's instructions.

5 Analytical determination

For analytical determination, 1 μL from each of the urine samples prepared according to Section 3.2 is injected into the gas chromatograph. The retention times of the IS, and the analytes EMC and NMF under the analytical conditions given in Section 4 are listed in Table 2. The retention times given in Table 2 are intended to be a rough guide only. Figure 4 shows a chromatogram of a processed urine sample obtained from an employee exposed to DMF.

Table 2 Retention times of the analytes and the IS under the given analytical conditions.

Analyte	Retention time [min]
DMPA (IS)	15.1
EMC	16.2
NMF	22.3

6 Calibration

The calibration standards (see Section 2.4) are prepared in the same way as the urine samples (see Section 3.2) and analysed according to Sections 4 and 5 using GC-NPD or GC-TSD. Calibration graphs are obtained by plotting the quotients of the peak areas of the analytes and the internal standard against the spiked concentration. It is not necessary to plot a complete calibration graph for every analytical series, because the inclusion of one calibration standard and one quality control sample (e.g. urine from a person exposed to DMF) within each analytical series is sufficient. New calibration graphs should be plotted if the quality control results indicate systematic deviations. The calibration graph is linear between the detection limits and between a concentration level of 100 mg AMCC or NMF per litre urine.

7 Calculation of the analytical result

To determine the analyte concentration in a sample, the peak area of the analyte is divided by the peak area of the internal standard DMPA. The obtained quotient is inserted into the corresponding calibration graph (see Section 6). Accordingly, the AMCC or NMF concentration in mg per litre urine is obtained. A reagent blank value (bidist. water) should be included with every analytical series, and if a value for the blank sample is observed, this value can be subtracted from the analytical result. Nevertheless, the contamination source should be identified and eliminated.

8 Standardisation and quality control

Quality control of the analytical results is carried out as stipulated in the guidelines of the *Bundesärztekammer* (German Medical Association) [11–13]. To check precision, a spiked urine sample with a constant and known concentration level of NMF and AMCC is analysed within each analytical series. As material for quality control is not commercially available, it must be prepared in the laboratory. For this purpose, a defined quantity of NMF and AMCC is added to pooled urine of persons occupationally non-exposed to DMF. The control material is divided into aliquots and stored in sealable 20 mL screw cap glass tubes at –20°C. The concentration level of the control material should lie within the relevant concentration range. The nominal value and the tolerance range of this quality control material are determined in a pre-analytical period [13].

9 Evaluation of the method

9.1 Precision

To determine the within day precision urine samples spiked with concentrations of 30 mg NMF and 40 mg AMCC per litre urine were consecutively analysed five times. Furthermore, the precision from day to day was determined by preparing and analysing the same sample material on five different days. The precision data obtained are listed in Table 3.

Table 3 Within day and day to day precision (n = 5).

Parameter	Analyte level [mg/L]	Standard deviation (rel.) [%]	Prognostic range [%]
Within day precision (n = 5)			
NMF	30	2.6	7.2
AMCC	40	9.4	26.1
Day to day precision (n = 5)			
NMF	30	5.3	14.7
AMCC	40	10.8	30.0

9.2 Accuracy

Recovery experiments with spiked urine samples were carried out to determine the accuracy of the method. For this purpose, pooled urine was spiked with defined analyte levels, processed eight times and analysed. The mean relative recovery rates thus obtained are shown in Table 4.

Table 4 Relative recovery rates and analyte losses due to sample preparation in spiked urine samples (n = 8).

Parameter	Spiked analyte level [mg/L]	Mean relative recovery [%]	Losses due to sample preparation [%]
NMF	25	94.6	21.5
AMCC	25	81.4	64.8

Furthermore, the losses due to sample preparation were determined. For this purpose, the peak areas of the spiked urine samples were compared with the peak areas of standards in organic solvents not subjected to any further processing and directly injected into the analytical system (in the case of AMCC the final product ethyl-N-methylcarbamate (EMC) was used as reference and a conversion rate of AMCC to EMC of 100% was assumed). The losses due to sample preparation are also shown in Table 4.

9.3 Detection limits

Under the given conditions the detection limits of the analytes were 1.0 mg/L urine for NMF and 0.5 mg/L urine for AMCC (based on a signal to noise ratio of 3:1). The resulting quantitation limits (signal to noise ratio of 9:1) are 3.0 mg/L urine for NMF and 1.5 mg/L urine for AMCC.

9.4 Sources of error

To obtain valid and reproducible results (especially for AMCC) careful attention must be paid to the following sources of error during sample preparation:

a) After shaking the samples with 1 N hydrochloric acid, the IS and sodium chloride, a residue of undissolved sodium chloride should remain at the bottom of the glass before the sample is extracted with THF. If this is not the case more sodium chloride must be added and the sample shaken again. This step must be carefully carried out to avoid a reduced yield of extracted AMCC and NMF or HMMF.

b) The THF used for extraction should not contain any stabilizer. For practical as well as technical safety reasons, THF flasks containing low quantities (maximum 500 mL) should therefore be stored in the laboratory, and if possible ordered immediately before analysis. Non-observance of this step may produce an increased background noise level and a loss of AMCC during sample preparation, especially when the THF – after first opening – is stored longer than two weeks.

c) Ensure that all sample tubes are tightly closed during shaking to avoid losses.

d) The samples may be heated slightly during all evaporation steps under a stream of nitrogen. The purpose of such heating should however only be to compensate for cooling through evaporation, and not for heating the samples. Evaporation to dryness should never take place. The amount of residue (e.g. 500 µL THF) can be estimated using a reference tube into which the corresponding amount of water has been pipetted.
e) The conversion of AMCC to ethyl-N-methylcarbamate (EMC) in the presence of ethanol and potassium carbonate demands exact preparation. There must be no delay as this would inevitably produce a loss of AMCC. Even under optimum conditions, conversion rates of only approx. 40% are attained. Therefore, all preparations for this step should be done while the THF is being reduced by evaporation. This preparation comprises filling of the 5 mL screw cap glass tubes with 1.5 g potassium carbonate, having the corresponding screw caps (with septa already inserted) nearby, together with a 5 mL pipette (to transfer the sample to the screw cap glass tube in one single step) and the laboratory shaker. After addition of the ethanolic solution to the potassium carbonate, the sample tube should be immediately sealed and shaken thoroughly. This step must be carried out separately for each sample, because serial addition of the ethanolic solutions to the potassium carbonate sample glasses, together with subsequent shaking of all samples may result in considerable analyte losses.

10 Discussion of the method

The method is especially suitable for simultaneous detection of AMCC as well as of the sum of HMMF and NMF in occupational medical practice [6], and is sufficiently sensitive for monitoring the BAT value of 35 mg NMF per litre urine in persons occupationally exposed to DMF. The type of method presented here is similar to the determination of HMMF and AMCC described by Mráz [14], with the exception of an additional extraction step using THF. A further difference is that Mráz already adds potassium carbonate to the urine samples at the beginning of sample preparation. This increases the pH value of the urine sample into the alkaline range and ensures early conversion of AMCC to EMC. It also ensures the conversion of HMMF to NMF so that NMF is present already upon injection into the gas chromatograph. The extraction of HMMF from urine samples using THF under the conditions given here is complete; therefore there is no difference between either method with regard to their quality criteria. Using a TSD detector, the detection limit of the method presented here is dependent on the quality of the detection bead. During examination of the method even lower detection limits were obtained than those described in Section 9.3 on account of better bead quality. If maximum sensitivity is necessary, a number of beads can be analysed in advance for their quality or an alternative extended sample preparation can be used.

For this purpose, 1 mL of the ethanolic supernatant (of the sample preparation for GC/NPD or GC/TSD) is transferred to a 20 mL screw cap glass tube. Subsequently, 3 mL bidistilled water are added and extraction is performed twice using 5 mL dichloromethane for each extraction. Subsequently, the samples are centrifuged at 2600 g for 5 min. The dichloromethane phases are combined in a new 20 mL screw cap glass tube and dried above anhydrous sodium sulphate for 30 min. The samples are then transferred to a new screw cap glass tube and the sodium sulphate is washed again with 3 mL dichloromethane. The combined organic phases are reduced by evaporation to about 100 µL under a weak stream of nitrogen and slight heating, and are then transferred to a microinsert already containing 50 µL ethanol. After the final evaporation to 50 µL sample solution, the sample is separated by capillary gas chromatography and the metabolites NMF and AMCC are detected using mass spectrometry (GC-MS). The detection limits of this sensitive but very comprehensive method are about ten times lower (0.1 mg/L for NMF, 0.03 mg/L for EMC) compared to the described GC/NPD- or GC/TSD-method. Conversely, the result is nearly the same when compared to further reliability data. The GC-MS method is suitable for the determination of AMCC background levels in human urine (e.g. produced by endogenous carbamoylating species) from persons occupationally non-exposed to DMF [15]. As part of the GC-MS process, the following mass traces are analysed: DMPA (m/z 101 and 72), EMC (m/z 103 and 75) and NMF (m/z 59 and 30). The respective higher mass traces are used for quantitation, whereas the lower mass traces and the respective ratio between both masses are used as quality criteria [16].

The method described here for the determination of HMMF and AMCC using GC/NPD or GC/TSD for occupational monitoring (or the extension of the method for AMCC for environmental monitoring using GC-MS) represents an extension of the method previously published by the DFG for the detection of HMMF. For the first time, two relevant metabolites of DMF can be detected simultaneously as part of a single analytical process.

11 References

1 J. Mráz and H. Nohová: Absorption, metabolism and elimination of N,N-dimethylformamide in humans. Int. Arch. Occup. Environ. Health 64, 85–92 (1992).

2 J. Mráz, H. Cross, A. Gescher, M.D. Threadgill, J. Flek: Differences between rodents and humans in the metabolic toxification of N,N-dimethylformamide. Toxicol. Appl. Pharmacol. 98, 507–516 (1989).

3 A. Gescher: Metabolism of N,N-dimethylformamide: key to the understanding of its toxicity. Chem. Res. Toxicol. 6, 245–251 (1993).

4 H.U. Käfferlein and J. Angerer: N-methylcarbamoylated valine of hemoglobin in humans after exposure to N,N-dimethylformamide: evidence for the formation of methyl isocyanate? Chem. Res. Toxicol. 14, 833–840 (2001).

5 T. Sakai, H. Kageyama, T. Araki, T. Yosida, T. Kuribayashi, Y. Masuyama: Biological monitoring of workers exposed to N,N-dimethylformamide by determination of the urinary me-

tabolites, N-methylformamide and N-acetyl-S-(N-methylcarbamoyl) cysteine. Int. Arch. Occup. Environ. Health 67, 125–129 (1995).
6   H.U. Käfferlein, T. Göen, J. Müller, R. Wrbitzky, J. Angerer: Biological monitoring of workers exposed to N,N-dimethylformamide in the synthetic fibre industry. Int. Arch. Occup. Environ. Health 73, 113–120 (2000).
7   M. Imbriani, L. Maestri, P. Marraccini, G. Saretto, A. Alessio, S. Negri, S. Ghittori: Urinary determination of N-acetyl-S-(N-methylcarbamoyl)cysteine and N-methylformamide in workers exposed to N,N-dimethylformamide. Int. Arch. Occup. Environ. Health 75, 445–452 (2002).
8   Deutsche Forschungsgemeinschaft (DFG): List of MAK and BAT values 2012. 48th issue. Wiley-VCH Verlag, Weinheim (2012).
9   H. Greim (ed.): Dimethylformamide. MAK Value Documentations. Volume 8, Wiley-VCH Verlag, Weinheim (1997).
10 A. Hartwig (ed.): Dimethylformamide. The MAK Collection Part I: MAK Value Documentations. Volume 26, Wiley-VCH Verlag, Weinheim (2010).
11 Bundesärztekammer: Qualitätssicherung der quantitativen Bestimmungen im Laboratorium. Neue Richtlinien der Bundesärztekammer. Dt. Ärztebl. 85, A699–A712 (1988).
12 Bundesärztekammer: Ergänzung der "Richtlinien der Bundesärztekammer zur Qualitätssicherung in medizinischen Laboratorien". Dt. Ärztebl. 91, C159–C161 (1994).
13 J. Angerer, T. Göen, G. Lehnert: Mindestanforderungen an die Qualität von umweltmedizinisch-toxikologischen Analysen. Umweltmed. Forsch. Prax. 3, 307–312 (1998).
14 J. Mráz: Gas chromatographic method for the determination of N-acetyl-S-(N-methylcarbamoyl)cysteine, a metabolite of N,N-dimethylformamide and N-methylformamide, in human urine. J. Chromatogr. 431, 361–368 (1988).
15 H.U. Käfferlein and J. Angerer: Determination of N-acetyl-S-(N-methylcarbamoyl)cysteine (AMCC) in the general population using gas chromatography-mass spectrometry. J. Environ. Monit. 1, 465–469 (1999).
16 H.U. Käfferlein and J. Angerer: Simultaneous determination of two human urinary metabolites of N,N-dimethylformamide using gas chromatography-thermionic sensitive detection with mass spectrometric confirmation. J. Chromatogr. B 734, 285–298 (1999).

Authors: *H. U. Käfferlein, J. Angerer*
Examiner: *G. Leng*

12 Appendix

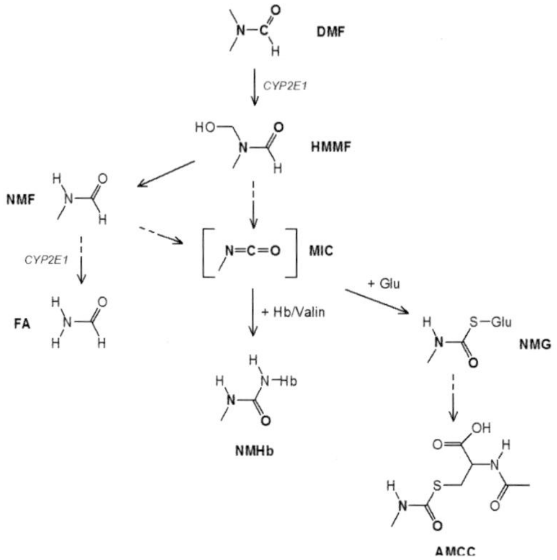

Fig. 1 Presently known metabolism of N,N-dimethylformamide. Abbreviations used: AMCC N-acetyl-S-(N-methylcarbamoyl)-cysteine; DMF N,N-dimethylformamide; FA formamide; HMMF N-hydroxymethyl-N-methylformamide; MIC methylisocyanate; NMF N-methylformamide; NMG N-methyl glutathione; NMHb N-methylcarbamoyl haemoglobin (at the N-terminal valine).

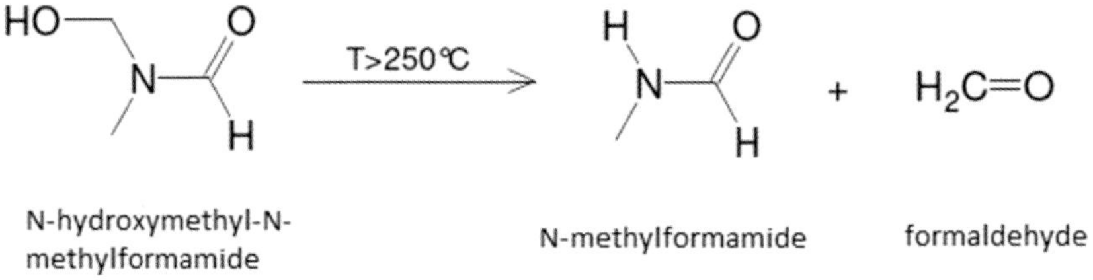

Fig. 2 Thermal decomposition of N-hydroxymethyl-N-methylformamide (HMMF) in the injector block of the gas chromatograph at temperatures ≥250°C to formaldehyde and N-methylformamide.

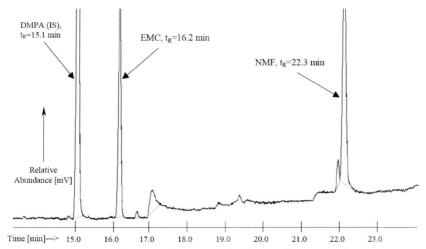

N-acetyl-S-(N-methylcarbamoyl) cysteine   ethyl-N-methylcarbamate

Fig. 3 Conversion of N-acetyl-S-(N-methylcarbamoyl)-cysteine (AMCC) to ethyl-N-methylcarbamate (EMC) in the presence of ethanol and potassium carbonate.

Fig. 4 Chromatogram of a processed urine sample of an employee exposed to DMF. Determined analyte concentration levels were: 34.4 mg/L AMCC (EMC, t_R = 16.2 min) and 25.1 mg/L HMMF (NMF, t_R = 22.3 min). Internal standard: DMPA (t_R = 15.1 min).

Arsenic (total) in urine

Matrix:	Urine
Hazardous substances:	Arsenic and its compounds
Analytical principle:	ICP-Quadrupole mass spectrometry (ICP-MS)
Completed in:	November 2007

Overview of parameters that can be determined with this method and the corresponding hazardous substances:

Hazardous substance	CAS	Parameter	CAS
Arsenic	7440-38-2		
Inorganic arsenic compounds	–	Arsenic	7440-38-2
Organic arsenic compounds	–		

Summary

The method describes the determination of the total arsenic in urine using inductively coupled plasma mass spectrometry (ICP-MS). The method is rapid, simple, reliable, very sensitive and suitable for routine analysis in clinical laboratories with high sample throughput. It is possible to determine arsenic in the occupational and environmental concentration range. The urine samples are prepared by 1:10 (v:v) dilution with nitric acid and addition of rhodium as internal standard. For quantitation, the standard addition method in urine is used, which is then transformed into an external calibration. Spectral interferences on the isotope $^{75}As^+$ (e.g. $^{40}Ar^{35}Cl^+$) can be removed by the use of collision or reaction cell technology of modern ICP-MS instruments.

Arsenic

Within day precision: Standard deviation (rel.) s_w = 1.9% or 2.0%
Prognostic range u = 4.2% or 4.5%
at a nominal concentration of 34 µg or 184 µg arsenic per litre urine and where n = 10 determinations

Day to day precision: Standard deviation (rel.) s_w = 6.2% or 6.6%
Prognostic range u = 12.9% or 13.8%
at a nominal concentration of 34 µg or 184 µg arsenic per litre urine and where n = 20 determinations

Accuracy: Recovery rate (rel.) r = 101%
at a nominal concentration of 22 µg arsenic per litre urine and where n = 5 determinations

Detection limit: 0.1 µg arsenic per litre urine
Quantitation limit: 0.3 µg arsenic per litre urine

Arsenic

Arsenic (As, relative atomic mass 74.92, atomic number 33) is ubiquitously found in the nature. It is the 20th most abundant element in the earth's crust and is mainly found in the form of arsenides, sulfides or oxides. The most common oxidation states for arsenic are +V, +III, 0 and –III. Arsenic compounds can be subdivided into inorganic, water soluble compounds (e.g. arsenic trioxide (As_2O_3), arsenic pentoxide (As_2O_5), arsenious acid (H_3AsO_3) or arsenic acid (H_3AsO_4)), poorly soluble or insoluble inorganic compounds (e.g. arsenic(III) sulfide (As_2S_3)), organic arsenic compounds (e.g. arsenobetaine, arsenosugar) and gaseous arsenic compounds (e.g. arsine (AsH_3)) [1].

In the past, arsenic was used in the pharmaceutical industry and as plant protection agent, but did not play a role in these areas for a long time due to toxicological reasons. Today, arsenic is newly discovered for the treatment of patients with leukaemia. It is also used in the semi-conductor manufacturing, as well as in the production of paints and glass. In these cases occupational exposure to arsenic may occur. Another source of exposure is the extraction of metals from arsenic-containing ores, as arsenic often occurs in copper, iron or nickel ores. Ore smelting and combustion of fossil fuels lead to an emission of arsenic into the environment.

Arsenic concentrations in soil or ground water are strongly influenced by geochemical conditions. In some countries, such as Bangladesh, India or Chile extremely high arsenic concentrations in the soil were found. An environmental disaster in Bangladesh became a matter of public knowledge after wells were drilled into the groundwater containing high arsenic concentrations, without taking into account the high level of arsenic in the obtained water for the population. This has dramatic health consequences for the local population up to the present day [2]. In Germany, the arsenic level in drinking water is typically below the legal limit of 10 µg/L [3, 4]. Nevertheless, some mineral waters – due to variations in the geological distribution of arsenic – contain more than 10 µg/L of arsenic. The dietary intake of arsenic is mostly associated with the consumption of sea fish and other marine animals, whereby arsenobetaine represents the largest amount of the ingested arsenic [5].

Arsenic and its inorganic compounds (arsenic trioxide, arsenic pentoxide, arsenious acid, arsenic acid as well as its salts) are classified into Category 1 for carcinogenic substances by the Senate Commission for the Investigation of Health Hazards of Chemical Compounds in the Work Area [6]. A comprehensive toxicological review of arsenic and its inorganic compounds can be found in the 2002 MAK Documentation [7]. For inorganic arsenic and its methylated metabolites in urine, a *"Biologischer Leitwert"* (BLW) of 50 µg arsenic per L urine [6, 8] and a *"Biologischer Arbeitsstoff-Referenzwert"* (BAR) of 15 µg arsenic per L urine was established [6, 9]. For arsenic trioxide, exposure equivalents for carcinogenic substances (EKA) were assessed (50 µg/L arsenic in urine at a concentration of 0.01 mg/m^3 in the air, 90 µg/L in urine at a concentration of 0.05 mg/m^3 in the air, and 130 µg/L in urine at a concentration 0.1 mg/m^3 in the air [6, 10]).

Methods for the determination of toxicologically relevant arsenic [11] and of arsenic species [12] have already been published in this collection. To determine the toxicologically relevant arsenic, the hydride AAS technique is frequently used [11–14]. This technique allows the determination of different arsenic species such as As (III) and As (V) as well as the methylated arsenic species, monomethylarsonic acid and dimethylarsinic acid (MMA and DMA) which all form the gaseous hydride, while important organic arsenic compounds (e.g. arsenobetaine, arsenocholine) are not determined by this method. However, in practice difficulties with calibration are encountered when using flow injection for hydride AAS, because not all arsenic species are determined with the same sensitivity. Furthermore, additional hydrid forming organic arsenic species, e.g. arsenosugars, have recently been discovered [15] and may wrongly increase the toxicological relevant arsenic concentration determined with the hydride generation AAS. This means that the determination of hydride-forming arsenic species is only suitable to a limited extent for the assessment of health risks. The quantitation of individual arsenic spe-

Table 1 Data on background levels of arsenic in urine.

Country/Region	Category of persons/background level	n	Ref.
Northern and Western Germany	Adults (n = 87): 1–375 µg/L; geom. mean: 13 µg/L; 95th percentile: 143 µg/L Children (n = 72): 1–260 µg/L; geom. mean: 12 µg/L; 95th percentile: 91 µg/L	159	[31]
Argentina	43–403 µg/L	10	[32]
Belgium	17–230 µg/g creatinine	15	[33]
EU	Adults, 0.5–197 µg/L; geom. mean: 14 µg/L; 95th percentile: 149 µg/L	63	[34]
Belgium	21.8-1650 µg/day	9	[35]
Korea	mean value (n = 10): 69 µg/L (1–10 years) mean value (n = 10): 157 µg/L (10–20 years) mean value (n = 10): 159 µg/L (20–30 years) mean value (n = 10): 210 µg/L (30–40 years) mean value (n = 10): 226 µg/L (50–60 years)	50	[36]

cies in urine using HPLC/ICP-MS [16–19] or HPLC/AAS [20–22] provides more information about the arsenic species. Data on the concentration of individual arsenic species in the urine of the general population [17–19, 23–27] as well as in the urine of occupationally exposed persons [28–30] have been published in the literature.

Literature data on background levels of total arsenic in urine are listed in Table 1. Data obtained by hydride AAS without sample digestion are not listed, as not all arsenic species form gaseous arsine without previous digestion of the urine.

Author: *P. Heitland*
Examiner: *B. Michalke*

Arsenic (total) in urine

Matrix:	Urine
Hazardous substances:	Arsenic and its compounds
Analytical principle:	ICP-Quadrupole mass spectrometry (ICP-MS)
Completed in:	November 2007

Contents

1 General principles
2 Equipment, chemicals and solutions
2.1 Equipment
2.2 Chemicals
2.3 Solutions
2.4 Calibration standards
3 Specimen collection and sample preparation
4 Operational parameters
4.1 Sample introduction and plasma settings
4.2 Operational parameters for mass spectrometry
5 Analytical determination
6 Calibration
7 Calculation of the analytical result
8 Standardisation and quality control
9 Evaluation of the method
9.1 Precision
9.2 Accuracy
9.3 Detection limit and quantitation limit
9.4 Sources of error
10 Discussion of the method
11 References
12 Appendix

1 General principles

Determination of arsenic is carried out using mass spectrometry with inductively coupled plasma as ion source (ICP-MS). The urine samples are diluted 1:10 (v:v) and are introduced directly into the hot plasma (ICP) by pneumatic nebulization. In the plasma the samples are evaporated, dissociated, atomized and ionized. The ions from the ICP are transported into the mass spectrometer and are separated based on their mass to charge ratio and detected using an electron multiplier.

Calibration solutions in urine are prepared to establish a calibration curve for the determination of total arsenic. A linear correlation between signal intensities and mass concentrations of total arsenic is obtained over several orders of magnitude. To improve accuracy and precision, rhodium is added to all samples as internal standard.

2 Equipment, chemicals and solutions

2.1 Equipment

- ICP-mass spectrometer (Quadrupole with reaction/collision gas cell or sector field unit) with autosampler (e.g. Agilent Technologies 7500)
- Stroke pipettes with adjustable volume between 10–100 µL or 100–1000 µL with suitable pipette tips (e.g. Eppendorf)
- 1000 mL Polyethylene flask with bottletop dispenser (continuously adjustable between 0.5 and 5 mL) (e.g. Dispensette™, Brand)
- 10 mL and 100 mL Volumetric flasks (e.g. Schott)
- 10 mL Polypropylene vials (e.g. Sarstedt)
- Laboratory shaker (e.g. Reax 2000; Heidolph Instruments GmbH)
- 250 mL Screw cap container for urine collection (e.g. Sarstedt No. 77.577)

2.2 Chemicals

- Nitric acid 65% Suprapur™ (e.g. Merck, No. 100441)
- Arsenic standard 1000 mg/L As (e.g. SPEX CertiPrep, PLAS2-2Y)
- Rhodium standard 10 mg/L Rh (e.g. Merck, 108525)
- Analytical quality control material (e.g. Seronorm™ Trace Element Urine)
- Deionised water
- Argon 99.996% for the plasma (e.g. Linde)
- Argon 99.999% for the collision/reaction cell (e.g. Linde)
- Helium 99.999% for the collision/reaction cell (e.g. Linde)

2.3 Solutions

- Solution of the internal standard (50 µg/L rhodium)
  50 mL deionised water and 1 mL concentrated nitric acid are placed in a 100 mL volumetric flask to which 500 µL of the 10 mg/L rhodium standard solution are added. The solution is then made up to the mark with deionised water and homogenised by shaking. The solution contains 50 µg/L rhodium and is freshly prepared every day.

2.4 Calibration standards

- Working solution 1 (1 mg/L arsenic):
  50 mL deionised water and 1 mL concentrated nitric acid are placed in a 100 mL volumetric flask to which 100 µL of the 1000 mg/L arsenic standard solution are added. The solution is then made up to the mark with deionised water and homogenised by shaking. The solution contains 1 mg/L arsenic and is freshly prepared every week.
- Working solution 2 (100 µg/L arsenic):
  5 mL deionised water and 100 µL concentrated nitric acid are placed in a 10 mL volumetric flask to which 1000 µL of the 1 mg/L arsenic working solution 1 are added. The solution is made up to the mark with deionised water and then homogenised by shaking. The solution contains 100 µg/L arsenic and is freshly prepared every day.

Calibration is performed in urine by standard addition procedure. Calibration standard solutions are prepared by adding the solution of the internal standard (IS), nitric acid, defined volumes from the arsenic working solutions 1 and 2, and deionised water to a human urine sample with a low arsenic background level (final dilution of the urine is 1:10 (v:v)). This standard addition calibration is then converted into an external calibration. The pipetting scheme for the preparation of the calibration standards is given in Table 2. Thus, for example, to prepare the 1 µg/L calibration standard, 500 µL of the solution of the internal standard, 100 µL nitric

Table 2 Pipetting scheme for the preparation of calibration standard solutions.

Urine [µL]	IS [µL]	Nitric acid [µL]	Working solution 1 [µL]	Working solution 2 [µL]	Final volume [µL]	Analyte level [µg/L]
500	500	100	–	–	5000	0
500	500	100	–	50	5000	1
500	500	100	–	250	5000	5
500	500	100	–	500	5000	10
500	500	100	100		5000	20
500	500	100	250		5000	50

acid and 50 µL of the working solution 2 and 500 µL of the urine are pipetted into a 10 mL plastic tube. After this, 3.85 mL deionised water are added giving a final volume of 5 mL. Then, the solution is thoroughly mixed on a laboratory shaker. The other calibration standards are prepared according to the pipetting scheme given in Table 2.

3 Specimen collection and sample preparation

Highest purity is required for the chemicals and the equipment used. The urine should be collected in polyethylene containers previously rinsed with 1% nitric acid or in containers guaranteed free of arsenic according to manufacturer's information.

For the determination of environmental exposure to arsenic, the use of a 24-hour urine is recommended. Nevertheless, spontaneous urine samples or first morning urines may also be used. In the case of long-term exposure at the workplace, specimen collection after several previous work shifts or at the end of a shift is recommended [6].

If the determination of arsenic is not carried out within 1 or 2 days after specimen collection, the urine sample should be acidified (1 mL concentrated nitric acid per 100 mL urine) and stored in the refrigerator at 4°C. For a longer storage period over weeks or months the urine samples should be deep frozen at –18°C.

For preparation, the urine sample is shaken thoroughly. 500 µL urine are pipetted into a 10 mL vial, to which 500 µL of the solution of the internal standard, 100 µL nitric acid and 3.9 mL deionised water are added. Blank samples, using bidist. water instead of urine, are treated in the same manner.

4 Operational parameters

4.1 Sample introduction and plasma settings

The following plasma settings are intended to serve as a rough guide only. The parameters must be optimised accordingly for every type of ICP-MS instrument. The setting of additional parameters may be necessary when equipment from other manufacturers is used.

Plasma performance:	1.5 kW
Gas flow:	Nebuliser gas 1.15 L/min argon
	Middle gas 1 L/min argon
	Outer gas 15 L/min argon
Injector tube (torch)	Inner diameter 2.5 mm

Sample introduction:	Peristaltic pump, performance 0.4 mL/min
Nebuliser:	Babington
Atomiser chamber:	Scott type (cooled to 5°C)

Generally, other nebulisers can be used for sample introduction. For method development, a Babington nebuliser was used, as it is practically insensible to particles from the urine and shows a high long-term stability in routine work. A somewhat wider injector tube diameter (2.5 mm) in the torch is also recommended leading to a better long-term stability.

4.2 Operational parameters for mass spectrometry

Due to different spectrometer designs, the settings for the mass spectrometer are depending on the type of device used and must always be optimised accordingly. Additional setting and parameter optimisation may be necessary when using spectrometers of other manufacturers. The following parameters are intended to be a rough guide only for the applied instrument in this case.

Collision gas flow:	4 mL/min helium
Integration time:	2 s
Lens voltages:	2.8 V (extract lens 1)
	–84 V (extract lens 2)
	–16 V (omega bias)
	0 V (omega lens)
	–74 V (cell exit)
	–74 V (quadrupole bias)
	–74 V (octopole bias)

5 Analytical determination

Careful optimisation of collision or reaction gas flow is of special importance, as the polyatomic interferences occurring on isotope $^{75}As^+$ from $^{40}Ar^{35}Cl^+$ or $^{38}Ar^{37}Cl^+$ should be eliminated. To check, if the interference on isotope $^{75}As^+$ is negligible an arsenic-free 1% hydrochloric acid solution can be analysed.

The urine samples diluted in a ratio of 1:10 (v:v) are introduced directly to the ICP and analysed by mass spectrometry. Each sample is determined three times and the mean value used for the calculation of the analytical result.

6 Calibration

The calibration solutions are prepared according to Section 2.4. The calibration graph is obtained by plotting the intensity [cps] of the ^{75}As$^+$ mass peak against the spiked concentration.

The calibration graph is linear between the detection limit and a concentration level of 50 µg arsenic per litre urine. A new calibration graph should be plotted if the quality control results indicate systematic deviations.

7 Calculation of the analytical result

To determine the analyte concentration in a sample, the measured intensity of the ^{75}As$^+$ mass peak is inserted into the corresponding calibration graph (see Section 6). The arsenic concentration in µg per litre urine is obtained. The basal arsenic level in the spiked urine, as determined by standard addition procedure, and reagent blank values have to be accounted for, where necessary. Normally, calculation of the analytical result is performed by the software of the spectrometer.

8 Standardisation and quality control

Quality control of the analytical results is carried out as stipulated in the guidelines of the *Bundesärztekammer* (German Medical Association) [37].

For the quantitation of arsenic in urine, control material of different manufacturers and also certified reference material is commercially available. To cover a wide concentration range, two or more control materials with different concentrations should be used. These should be analysed after calibration, after every twentieth sample and at the end of each analytical series.

9 Evaluation of the method

9.1 Precision

To determine the within day precision arsenic levels in two control materials with nominal values of 34 µg/L or 184 µg/L arsenic in urine were analysed 10 times in a row. The relative standard deviations with the corresponding prognostic ranges are given in Table 3.

In addition, the day to day precision was determined. To do this, the same control materials were prepared and analysed on twenty different days as described in the preceding section. Table 4 gives the resultant precision data.

Table 3 Within day precision for the determination of arsenic in urine (n = 10).

Analyte	Nominal concentration [µg/L]	Standard deviation (rel.) s_w [%]	Prognostic range u [%]
Arsenic	34	1.9	4.2
	184	2.0	4.5

Table 4 Day to day precision for the determination of arsenic in urine (n = 20).

Analyte	Nominal concentration [µg/L]	Standard deviation (rel.) s_w [%]	Prognostic range u [%]
Arsenic	34	6.2	12.9
	184	6.6	13.8

9.2 Accuracy

Accuracy was determined by internal and external quality assurance. For internal quality assurance, control materials of different manufacturers were analysed. Table 5 shows the results. For some of the control materials, the measured concentrations are very close to the nominal values (Seronorm urine). The maximum deviation from the nominal value is 10% (Medisafe urine).

Additionally, a urine sample with a background level of 12 µg arsenic per litre was spiked with 10 µg arsenic per litre and analysed five times in order to determine the recovery rate. A mean recovery rate of 101% was obtained.

External quality assessment was done by participation in two international schemes. The first was the German External Quality Assessment Scheme (G-EQUAS) No. 37 for toxicological analyses in occupational and environmental medicine, carried out by the *Institut und Poliklinik für Arbeits-, Sozial- und Umweltmedizin* (IPASUM; Institute and Polyclinic for Occupational, Social and Environmental Medicine) at the University of Erlangen-Nuremberg in May 2006 [38]. The

Table 5 Results of internal quality assurance: determination of arsenic in urine.

Control material	Analysed value [µg/L]	Nominal value [µg/L]	Recovery rate r [%]
Recipe ClinChek urine	34	37	91.9
	73	69	106
Sero Seronorm urine	184	183	101
Biorad Lyphochek urine	71	76	93.4
Medichem Medisafe urine	50	55	90.9

second was the ICP-MS Reference Program for Biological Materials at the *Institut National De Santé Publique du Quebec* (INSPQ; Quebec National Institute for Public Health) in March 2006 (QMEQAS 06-2) [39]. In the first quality assessment scheme, the results obtained using this method were 41 and 330 µg/L arsenic at nominal values of 39 and 319 µg/L, respectively. In the second scheme the determined value was 72 µg arsenic per litre at a nominal value of 76.9 µg/L. The maximum deviation from the nominal value was thus below 7%. It was possible to meet the tolerance ranges of both external quality assessment schemes.

9.3   Detection limit and quantitation limit

The detection limit was determined from the signal to noise ratio and the standard deviation (n = 10) of the spectral background level based on the 3 s criterion [40]. Accordingly, a detection limit of 0.1 µg arsenic per litre urine in an undiluted sample was determined. This corresponds to a quantitation limit of 0.3 µg arsenic per litre urine.

9.4   Sources of error

The determination of ^{75}As$^+$ in urine using ICP-MS was for a long time problematic or impossible. The most common way to correct for the ^{40}Ar^{35}Cl$^+$ interference was the use of mathematical correction equations. The introduction of quadrupole units with collision/reaction cells or high resolution sector field units made the determination of arsenic with ICP-MS also possible. In the case of quadrupole ICP-MS, the interfering ^{40}Ar^{35}Cl$^+$ is eliminated as it dissociates or reacts through collision or reaction with a gas prior to entering the mass spectrometer. Using high resolution sector field ICP-MS, the interference is negligible, when the mass peaks of analyte and interfering ions are separated. During method development, helium was used as collision gas in an Agilent ICP-MS unit. Effective elimination of the interference was here possible after optimisation of the collision gas flow (optimum at 4 mL/min).

Interferences due to matrix compounds (organic compounds, salts, easily ionized elements, etc.) are compensated for by the 1:10 (v:v) dilution of the urine as well as by using rhodium as internal standard. Frequently, for urine analysis, digestion is recommended, whereby this has the disadvantage of being costly and time-consuming and additionally increases the risk of systematic errors.

But also without digestion, contaminations must be considered as a possible source of error, so that the cleanliness of the equipment and the purity of the chemicals used must always be ensured.

Analytical problems associated with urine analysis may occur due to blocked nebuliser capillaries or particle deposition in the injector tube of the torch. These problems can be avoided by the 1:10 dilution of the urine samples as well as by using a Babington nebuliser and an injector tube with a somewhat wider diameter. The choice of the injector tube diameter depends on the ICP instrument. The diameter should be wider than 1.8 mm when the urine is diluted 1:10.

Prior to diluting the sample, thorough shaking is recommended in order to homogenise the sample and to avoid adsorption of the analyte on sediment particles. ^{103}Rh was found to be a suitable internal standard. Rhodium is stable in the solutions and normally not found in real urine samples.

10   Discussion of the method

The present method allows a rapid, simple, reliable and sufficiently sensitive determination of total arsenic in urine. Preparation of 50 samples, including calibration standards and control samples could be done in a routine laboratory by a qualified lab assistant in about an hour. Control materials and certified reference materials for internal quality control are commercially available. Additionally, national and international quality assessment schemes are carried out several times a year [38, 39].

A detection limit of 0.1 µg/L permits the determination of arsenic in the occupational and environmental concentration range, as studies for background exposure have shown that even in non-exposed persons arsenic levels are rarely below 1 µg total arsenic per litre urine [31, 33, 34]. The ICP-MS method fulfills the requirements on precision and accuracy because of good precision data and recovery rates obtained from internal and external quality assessment. The method allows the determination of total arsenic due to the high plasma temperature, which converts all arsenic species into the ionised form. This can be confirmed by the fact that the slope of the calibration graphs is the same, no matter what type of arsenic species is used for calibration.

When determining low concentrations of total arsenic, an exposure to toxic arsenic species can be excluded. Only when the arsenic concentrations are high further speciation is required to get a toxicological conclusion. This is possible by coupling HPLC with ICP-MS [41] or AAS [12], which has already been published in this method collection.

Apart from the less time-consuming sample preparation and the high sensitivity, the present ICP-MS method has the additional advantage that other elements can be determined simultaneously. Furthermore, the procedure described here can be used as a reference method for species analysis.

Instruments used:

- ICP Mass spectrometer 7500ce with collision/reaction gas cell (Agilent Technologies) and ASX-520 autosampler (Cetac).

11 References

1 T. Gebel and H. Becher: VI-3 Metalle/Arsen. In: Handbuch der Umweltmedizin. 21. Ergänzung S. 1–25. Ecomed Medizin, Verlagsgruppe Hüthig Jehle Rehm GmbH, Heidelberg (2001).
2 A. Mushtaque and R. Chowdhury: Die Arsenkatastrophe von Bangladesh. Spektrum der Wissenschaft, 74–79 (2004).
3 Bundesgesetzblatt: Verordnung des Bundesministers für soziale Sicherheit und Generationen über die Qualität von Wasser für den menschlichen Gebrauch (Trinkwasserverordnung – TWV); erste Verordnung zur Änderung der Trinkwasserverordnung vom 03. Mai 2011. BGBl. II No. 21, 748–774 (2011).
4 Stellungnahme der Kommission "Human-Biomonitoring" des Umweltbundesamtes: Stoffmonographie Arsen – Referenzwert für Urin. Umweltmed. Forsch. Prax. 9, 313–322 (2004).
5 R. Heinrich-Ramm, S. Mindt-Prüfert, D. Szadkowski: Arsenic species excretion after controlled seafood consumption. J. Chromatogr. B 778, 263–273 (2002).
6 A. Hartwig (ed.): Deutsche Forschungsgemeinschaft, List of MAK and BAT Values 2012. Report No. 48. Wiley-VCH, Weinheim (2012).
7 H. Greim (ed.): Arsenic and its inorganic compounds (with the exception of arsine). The MAK-Collection Part I: MAK Value Documentations. Volume 21, Wiley-VCH, Weinheim (2005).
8 H. Greim and G. Lehnert (eds.): Arsenic trioxide. The MAK-Collection Part II: BAT Value Documentations. Volume 2, Wiley-VCH, Weinheim (1995).
9 H. Greim and H. Drexler (eds.): Arsenic and inorganic arsenic compounds (with the exception of arsenic hydride and its salts). The MAK-Collection Part II: BAT Value Documentations., Volume 4, Wiley-VCH, Weinheim (2005).
10 A. Hartwig and H. Drexler (eds.): Addendum zu Arsen und anorganische Arsenverbindungen (mit Ausnahme von Arsenwasserstoff). Biologische Arbeitsstoff-Toleranz-Werte (BAT-Werte) und Expositionsäquivalente für krebserzeugende Arbeitsstoffe (EKA) und Biologische Leitwerte (BLW), 18th issue, Wiley-VCH, Weinheim (2011).
11 D. Henschler, J. Angerer, K.H. Schaller (eds.): Arsenic. Analyses of Hazardous Substances in Biological Materials. Volume 3, VCH, Weinheim (1991).
12 H. Greim, J. Angerer, K.H. Schaller (eds.): Arsenic Species (As(III), As(V), monomethylarsonic acid, dimethylarsinic acid). In: Analyses of Hazardous Substances in Biological Materials. Volume 7, Wiley-VCH, Weinheim (2000).
13 V. Spevácová, M. Cejchanová, M. Cerná, V. Spevácek, J. Šmíd, B. Beneš: Population based biomonitoring in the Czech Republic: urinary arsenic. J. Environ. Monit. 4, 796–798 (2002).
14 T. Guo, J. Baasner, D.L. Tsalev: Fast automated determination of toxicologically relevant arsenic in urine by flow injection hydride generation atomic absorption spectrometry. Anal. Chim. Acta 349, 313–318 (1997).
15 E. Schmeisser, W. Goessler, N. Kienzl, K.A. Francesconi: Arsenosugars do form volatile analytes: Determination by HPLC-HG-ICPMS and implications for arsenic speciation analyses. Anal. Chem. 76, 418–423 (2004).
16 J.L. Sloth, E.H. Larsen, K. Julshamn: Selective arsenic speciation of human urine reference materials using gradient elution ion-exchange HPLC-ICP-MS. J. Anal. At. Spectrom. 19, 973–978 (2004).
17 J. Lintschinger, P. Schramel, A. Hatalak-Rauscher, I. Wendler, B. Michalke: A new method for the analysis of arsenic species in urine by using HPLC-ICP-MS. Fresenius J. Anal. Chem. 362, 313–318 (1998).

18 P. Heitland and H.D. Köster: Comparison of different medical cases in urinary arsenic speciation by fast HPLC-ICP-MS. Int. J. Hyg. Environ. Health 212, 432–438 (2009).
19 L.S. Milstein, A. Essader, E.D. Pellizarri, R.A. Fernando, J.R. Raymer, K.E. Levine, O. Akinbo: Development and application of a robust speciation method for determination of six arsenic compounds present in human urine. Environ. Health Perspectives 111, 293–296 (2003).
20 R. Sur, J. Begerow, L. Dunemann: Determination of arsenic species in human urine using HPLC with on-line photooxidation or microwave-assisted oxidation combined with flow-injection HG-AAS. Fresenius J. Anal. Chem. 363, 526–530 (1999).
21 M. Gómez, C. Cámara, M.A. Palacios, A. López-Gonzálvez: Anionic cartridge preconcentrators for inorganic arsenic, monomethylarsonate and dimethylarsinate determination by on-line HPLC-HG-AAS. Fresenius J. Anal. Chem. 257, 844–849 (1997).
22 R. Cornelis, X. Zhang, L. Mees, J. M. Christensen, K. Byrialsen, C. Dyrschel: Speciation measurements by HPLC-HGAAS of dimethylarsinic acid and arsenobetaine in three candidate lyphilized urine refrence materials. Analyst 123, 2883–2886 (1998).
23 D. Heitkemper, J. Creed, J. Caruso, F.L. Fricke: Speciation of arsenic in urine using high-performance liquid chromatography with inductively coupled plasma mass spectrometric detection. J. Anal. At. Spectrom. 4, 279–284 (1989).
24 V. Foa, A. Colombi, M. Maroni, M. Buratti, G. Calzaferri: The speciation of the chemical forms of arsenic in the biological monitoring of exposure to inorganic arsenic. Sci. Total Environ. 34, 241–259 (1984).
25 P. Kurttio, H. Komulainen, E. Hakala, H. Kahelin, J. Pekkanen: Urinary excretion of arsenic species after exposure to arsenic present in drinking water. Arch. Environ. Contam. Toxicol. 34, 297–305 (1998).
26 D.A. Kalman, J. Hughes, G. van Belle, T. Burbacher, D. Bolgiano, K. Coble, N.K. Mottet, L. Polissar: The effect of variable environmental arsenic contamination on urinary concentrations of arsenic species. Environ. Health Perspect. 89, 145–151 (1990).
27 R. Heinrich-Ramm, S. Mindt-Prüfert, D. Szadkowski: Arsenic species excretion in a group of persons in northern Germany – contribution to the evaluation of reference values. Int. J. Hyg. Environ. Health 203, 475–477 (2001).
28 M.W. Arbouine and H.K. Wilson: The effect of seafood consumption on the assessment of occupational exposure to arsenic by urinary arsenic speciation measurements. J. Trace Elem. Electrolytes Health Dis. 6, 153–160 (1992).
29 E. Hakala and L. Pyy: Selective determination of toxicologically important arsenic species in urine by high-performance liquid chromatography-hydride generation atomic absorption spectrometry. J. Anal. At. Spectrom. 7, 191–196 (1992).
30 J.C. Ng, D. Johnson, P. Imray, B. Chiswell, M.R. Moore: Speciation of arsenic metabolites in the urine of occupational workers and experimental rats using an optimised hydride cold-trapping method. Analyst 123, 929–933 (1998).
31 P. Heitland and H.D. Köster: Biomonitoring of 30 trace elements in urine of children and adults by ICP-MS. Clin. Chim. Acta 365, 310–318 (2006).
32 L.M. Del Razo, C. Aguilar, A. Sierra-Santoyo, M.E. Cebrián: Interference in the quantitation of methylated arsenic species in human urine. J. Anal. Tox. 23, 103–107 (1999).
33 J.F. Heilier, J.P. Buchet, V. Haufroid, D. Lison: Comparison of atomic absorption and fluorescence spectroscopic methods for the routine determination of urinary arsenic. Int. Arch. Occup. Environ. Health 78, 51–59 (2005).
34 P. Heitland and H.D. Köster: Fast, simple and reliable routine determination of 23 elements in urine by ICP-MS. J. Anal. At. Spectrom. 19, 1552–1558 (2004).

35 R. Cornelis, A. Speecke, J. Hoste: Neutron activation analysis for bulk and trace elements in urine. Anal. Chim. Acta 78, 317–327 (1975).

36 S.H. Nam, J.J. Kim, S.S. Han: Direct determination of total arsenic and arsenic species by ion chromatography coupled with inductively coupled plasma mass spectrometry. Bull. Korean Chem. Soc. 24, 1805–1809 (2003).

37 Bundesärztekammer: Richtlinie der Bundesärztekammer zur Qualitätssicherung quantitativer laboratoriumsmedizinischer Untersuchungen. Dt. Ärztebl. 105, A341–A355 (2008).

38 The German External Quality Assessment Scheme (G-EQUAS) No. 37 for Analyses in Biological Materials. Institute and Outpatient Clinic of Occupational, Social and Environmental Medicine of the University of Erlangen-Nuremberg, Germany (2006).

39 Quebec Multielement External Quality Assessment Scheme (QMEQAS), Institut National De Sante Publique Du Quebec (2006).

40 P.W.J.M. Boumans: Measuring detection limits in inductively coupled plasma emission spectrometry using the SBR-RSDB approach. I. A tutorial discussion of the theory. Spectrochim. Acta 46B, 431–45 (1991).

41 P. Heitland and H.D. Köster: Fast determination of arsenic species and total arsenic in urine by HPLC-ICP-MS: Concentration ranges of unexposed German inhabitants and clinical case studies. J. Anal. Toxicol. 32, 308–314 (2008).

Author: *P. Heitland*
Examiner: *B. Michalke*

12 Appendix

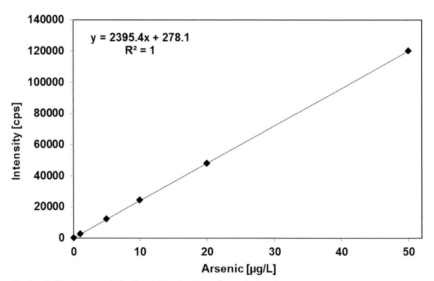

Fig. 1 Calibration graph for the determination of arsenic in urine.

Bisphenol A, genistein, daidzein and equol in urine

Matrix:	Urine
Hazardous substances:	Bisphenol A, genistein, daidzein
Analytical principle:	Capillary gas chromatography with mass selective detection (GC-MS)
Completed in:	November 2007

Overview of the parameters that can be determined with this method and the corresponding chemical substances:

Hazardous substance	CAS	Parameter	CAS
Genistein	446-72-0	Genistein	446-72-0
Daidzein	486-66-8	Daidzein	486-66-8
		Equol	531-95-3
Bisphenol A	80-05-7	Bisphenol A	80-05-7

Summary

The present method permits the quantitation of the isoflavones genistein and daidzein, of the daidzein metabolite equol and of bisphenol A in human urine.

For sample preparation, the internal standards are initially added to the urine samples. The analyte conjugates are then enzymatically hydrolysed, cleaned and enriched by means of solid phase extraction. After conversion to silyl ether derivatives the analytes are separated by gas chromatography and subjected to mass selective detection.

Genistein (GEN)

Within day precision: Standard deviation (rel.) s_w = 17% or 3%
Prognostic range u = 54% or 10%
at a spiked concentration of 10 or 125 µg GEN per litre urine and where n = 4 determinations

Day to day precision:	Standard deviation (rel.)	$s_w = 18\%$
	Prognostic range	$u = 46\%$
	at a spiked concentration of 75 µg GEN per litre urine and where n = 6 determinations	
Accuracy:	Recovery rate	$r = 114\%$ or 105%
	at a nominal concentration of 125 µg GEN per litre urine and where n = 4 determinations within day or from day to day	
Detection limit:	5 µg GEN per litre urine	
Quantitation limit:	18 µg GEN per litre urine	

Daidzein (DAI)

Within day precision:	Standard deviation (rel.)	$s_w = 5\%$ or 7%
	Prognostic range	$u = 16\%$ or 22%
	at a spiked concentration of 10 or 125 µg DAI per litre urine and where n = 4 determinations	
Day to day precision:	Standard deviation (rel.)	$s_w = 9\%$
	Prognostic range	$u = 23\%$
	at a spiked concentration of 75 µg DAI per litre urine and where n = 6 determinations	
Accuracy:	Recovery rate	$r = 109\%$ or 104%
	at a nominal concentration of 125 µg DAI per litre urine and where n = 4 determinations within day or from day to day	
Detection limit:	4 µg DAI per litre urine	
Quantitation limit:	9 µg DAI per litre urine	

Equol (EQ)

Within day precision:	Standard deviation (rel.)	$s_w = 6\%$ or 5%
	Prognostic range	$u = 19\%$ or 16%
	at a spiked concentration of 10 or 125 µg EQ per litre urine and where n = 4 determinations	
Day to day precision:	Standard deviation (rel.)	$s_w = 15\%$
	Prognostic range	$u = 39\%$
	at a spiked concentration of 75 µg EQ per litre urine and where n = 6 determinations	
Accuracy:	Recovery rate	$r = 89\%$ or 112%
	at a nominal concentration of 125 µg EQ per litre urine and where n = 4 determinations within day or from day to day	
Detection limit:	4 µg EQ per litre urine	
Quantitation limit:	11 µg EQ per litre urine	

Bisphenol A (BPA)

Within day precision:	Standard deviation (rel.)	$s_w = 7\%$ or 5%
	Prognostic range	$u = 22\%$ or 16%
	at a spiked concentration of 2.2 or 27.5 µg BPA per litre urine and where n = 4 determinations	
Day to day precision:	Standard deviation (rel.)	$s_w = 10\%$
	Prognostic range	$u = 26\%$
	at a spiked concentration of 16.5 µg BPA per litre urine and where n = 6 determinations	
Accuracy:	Recovery rate	$r = 94\%$ or 90%
	at a nominal concentration of 27.5 µg BPA per litre urine and where n = 4 determinations within day or from day to day	
Detection limit:	3 µg BPA per litre urine	
Quantitation limit:	7 µg BPA per litre urine	

Substances with estrogenic activity (genistein, daidzein, equol and bisphenol A)

Hormonally active substances or so-called "endocrine disruptors" are increasingly under discussion because exposure to these substances, especially in early life, can be accompanied by developmental and reproductive toxicity [1, 2]. In humans exposure usually takes place via food intake involving hormonally active substances of natural or anthropogenic origin [3, 4].

Phytoestrogens

Phytoestrogens are secondary plant ingredients including among other substances flavanones, flavones and isoflavones (see Figure 6 in the Appendix for basic structures). The best known representatives include the isoflavones daidzein (DAI) and genistein (GEN), which occur in many plants, but especially in soybean products [5]. Also the known phytoestrogens formononetin and biochanin A are enzymatically broken down to DAI and GEN (demethylation) by intestinal bacteria and liver microsomes [6, 7]. An overview of the phytoestrogen levels contained in foods is shown in Table 1.

The intake of DAI and GEN depends to a great extent on eating habits. The estimated daily intake of isoflavones is between 0.8 mg for a western style diet and 47 mg for a traditional Asian diet [14–17]. In more recent times, soy preparations are also gaining interest as dietary supplement providing relief in climacterium in the context of hormone replacement therapy. In such cases, isoflavone doses of 40 to 118 mg per day are recommended [18]. Through the enzyme systems in the gut flora of rodents and humans DAI is partially reduced to equol. Similarly, GEN is metabolised to 4-ethylphenol [19]. On account of its demonstrated xenoestrogenic structure, and as a metabolite of the isoflavone DAI, equol (EQ) has also been included in the method described below (see Figure 1 for structures).

Table 1 Phytoestrogen contents in foods.

Foods	Daidzein [µg/g]	Genistein [µg/g]	Formononetin [µg/g]	Biochanin A [µg/g]	Ref.
Alfalfa shoot	n.d.	n.d.	traces	traces	[8]
Bananas	n.d.	n.d.	–	–	[9]
Peas (green)	73	n.d.	traces	n.d.	[10]
Strawberries, fresh	0.045	0.457	–	–	[11]
Peanuts, roasted	0.037	0.172	–	–	[11]
Clover shoot	n.d.	4	22.8	4.4	[10]
Kidney beans	n.d.	n.d.	n.d.	4.1	[10]
Oranges	n.d.	n.d.	–	–	[11]
Beer (Pilsener Urquell)	0.0006	0.0018	0.004	0.0013	[12]
Soybeans	676–1007	612–1382	n.d.	n.d.	[10]
Soy protein, concentrate, washed in water	167–910	400–760	–	–	[13]
Soy shoot	138–225	113–305	–	–	[13]
Tofu	113	166	n.d.	n.d.	[10]

n. d. = not detected.

Bisphenol A

Bisphenol A (BPA) is a synthetic substance with endocrine activity and represents an important monomer in the production of epoxy resins, polycarbonates and other plastics. In 2009 the world production of BPA was 2.7 million tons, of which about 100 tons are released into the atmosphere per year [20]. This large-scale chemical product presents a hazard following exposure both in the production of the substance itself as well as in its further processing to polymers (polycarbonates and epoxy resins). On account of the endocrine activity of the substance a MAK value of 5 mg/m^3 (measured as inhalable fraction) was established for BPA [21]. Furthermore, a BLW (Biologischer Leitwert) for bisphenol A of 80 mg per litre urine has been established in 2006 [22].

BPA is found in many plastic flasks as well as in coatings of cans containing different foods. Through migration of BPA residual monomers into packed foods, the presence of BPA as contaminant in different beverages and foods can be demonstrated [23–25]. BPA has also been detected in the saliva of patients with tooth fillings in which the BPA monomer was used. The average exposure of an adult to BPA is about 30 µg per day (corresponding to approx. 0.5 µg/kg body weight) according to the European Commission [25].

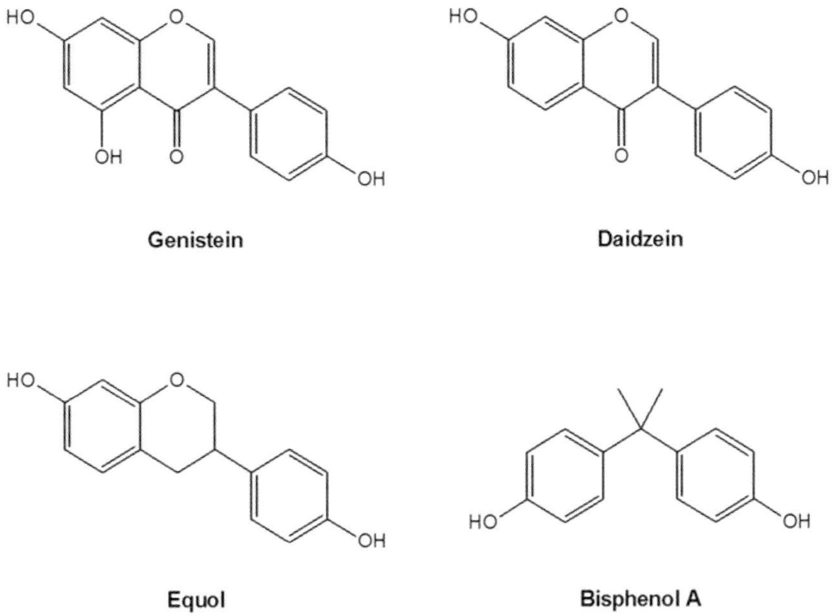

Fig. 1 Structures of the analysed substances with estrogenic activity.

These data form a frame of reference for the estimated daily intake of the estrogenically active isoflavones and BPA. However, as regards type and quantity, exposure can vary greatly. Although these exposure data form the basis for risk estimation, both the toxicodynamics (mode of action, hormonal activity) and the toxicokinetics (absorption, distribution, metabolism, elimination) as well as aspects of the affected endocrine system must also be taken into account for a toxicological evaluation of exposure to xenoestrogens [1, 26, 27]. The estrogenic activity of DAI, GEN and BPA was already analysed both *in vivo* and *in vitro* in many bioassays and was categorised as weak in comparison with the steroid estrogens [28–31]. Metabolism studies in rodents showed an extensive conjugation of isoflavones and BPA to polar metabolites by phase II enzymes [32–36]. In human studies a very similar metabolic pathway was observed [27, 37], which resulted in an effective elimination of the investigated xenoestrogens with elimination half-lives of 5-8 hours for DAI and GEN and less than 6 hours for BPA [38]. The compounds are mainly eliminated in urine in the form of polar metabolites. Exposure assessments are therefore possible from urinary biomonitoring studies. In the US-population, average urinary background levels of 68 µg DAI/g creatinine, 22 µg GEN/g creatinine and 8 µg EQ/g creatinine were found [39, 40].

The urinary analyte levels can provide useful information concerning external and internal exposure [34]. Whereas the lipophilic BPA is directly absorbed in the small intestine, the absorption of the analysed isoflavones is apparently dependent on the form in which they are present in the food (aglycon or conjugate) [41, 42]. In soybeans, DAI and GEN are mainly available as glucosides, acetyl glucosides or malonyl glucosides, which must be deconjugated before absorption. The sugar residues can be cleaved by gastric acid, β-glucosidases and/or bacterial enzymes in the intestine [43]. Figure 2 shows the main pathways of isoflavone metabolism via phase II enzymes to glucuronides, sulfates or sulfato-glucuronides. The conjugates are rapidly eliminated with the urine.

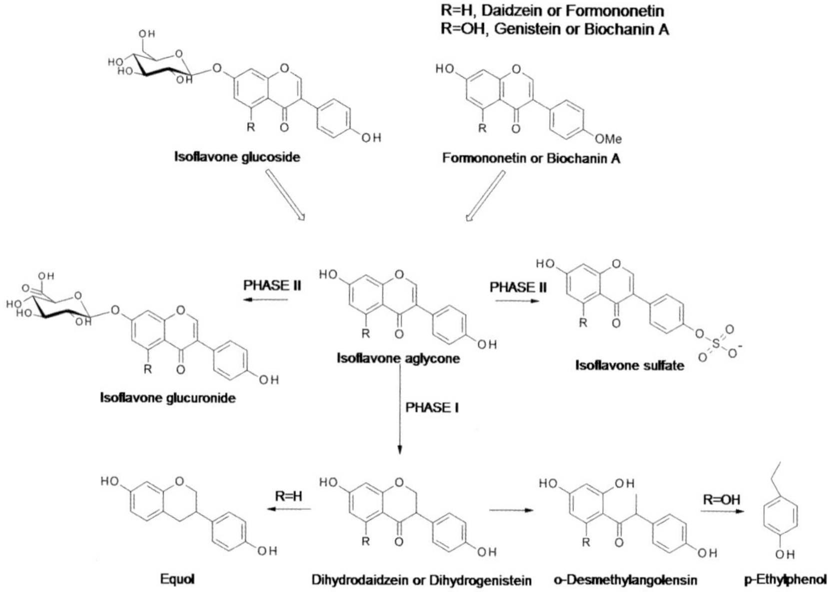

Fig. 2 Metabolism of isoflavones (only the main metabolites are shown).

Concerning hormonal activity, the conjugates of the isoflavones and of BPA are of minor importance, as they do not bind sufficiently to the intracellular estrogen receptor [35, 44–46]. In contrast, EQ shows *in vitro* an estrogenic activity similar to or even greater than that of DAI [47, 48]. 30–50% of the population are so-called equol eliminators, who are able to metabolise DAI to EQ [17, 42, 49, 50]. It is for this reason that EQ should always be determined when analysing isoflavones in the urine.

Authors: S. Moors, M. Blaszkewicz
Examiner: D. Barr

Bisphenol A, genistein, daidzein and equol in urine

Matrix:	Urine
Hazardous substances:	Bisphenol A, genistein, daidzein
Analytical principle:	Capillary gas chromatography with mass selective detection (GC-MS)
Completed in:	November 2007

Contents

1	General principles
2	Equipment, chemicals and solutions
2.1	Equipment and material
2.2	Chemicals
2.3	Solutions
2.4	Internal standards
2.5	Calibration standards
3	Specimen collection and sample preparation
3.1	Specimen collection
3.2	Sample preparation
4	Operational parameters
4.1	Operational parameters for gas chromatography
4.2	Mass spectrometric parameters
5	Analytical determination
6	Calibration
7	Calculation of the analytical result
8	Standardisation and quality control
9	Evaluation of the method
9.1	Precision
9.2	Accuracy
9.3	Detection limits and quantitation limits
9.4	Validation data of the alternative method
9.5	Sources of error
10	Discussion of the method

11 References
12 Appendix

1 General principles

To determine the isoflavones genistein and daidzein, the daidzein metabolite equol and bisphenol A in human urine samples, the internal standard substances are initially added to the samples. Then the urinary analyte conjugates are enzymatically hydrolysed, cleaned up and enriched using solid phase extraction. After conversion to silyl ether derivatives the analytes are separated by gas chromatography and subjected to mass selective detection.

2 Equipment, chemicals and solutions

2.1 Equipment and material

- Gas chromatograph (GC) with mass selective detector (MS), autosampler and data processing system (e.g. Agilent 6890N, Quadrupol MS 5973N, autosampler 7683, Software: Chemstation)
- Cooled injection system (e.g. Gerstel CIS 4 plus with Controller 505, Software: Master-software V1.8x)
- Gas chromatographic column: length: 15 m, inner diameter: 0.25 mm, stationary phase: 100% dimethylpolysiloxane, film thickness: 0.25 µm (e.g. DB-1-MS of J&W Scientific)
- pH meter (e.g. Metrohm)
- Thermostatic shaker (e.g. Thermomixer Comfort by Eppendorf)
- Vacuum station for solid phase extraction and dryer top (e.g. Phenomenex)
- Vacuum centrifuge (e.g. SPD 111V Speedvac, RVT 400 of Thermo Savant)
- Cooled vacuum trap (e.g. VN 100 vacuum Controller byThermo Electron)
- Membrane vacuum pump (e.g. MZ2C by Vaccubrand)
- Ultrasonic bath (e.g. Sonorex Rk 102H by Bandelin)
- Vortex mixer (e.g. Heidolph)
- Centrifuge (e.g. Eppendorf)
- Eppendorf tubes 0.5 mL, 1.5 mL and 2.0 mL (e.g. Eppendorf)
- Various glass beakers and glass flasks (e.g. Brand, Schott)
- C18 solid phase columns (e.g. Empore Extraction Cartrigde 4115 SD&HD of 3M)
- Lichrolut® RP 18 solid phase columns (500 mg, 3 mL polypropylene cartridges, e.g. Merck)
- Various volumetric flasks (e.g. Brand)
- Various pipettes with variable volumes (e.g. Eppendorf, Gilson)
- Pipette tips (e.g. Eppendorf, Sarstedt)
- 2 mL Sample vials made of borosilicate glass with 0.2 mL borosilicate glass in-

serts with polymer base, screw caps and silicon/polytetrafluoroethylene (PTFE) septa (e.g. VWR)
- Microlitre syringes 10 μL and 25 μL (e.g. Hamilton, SGE)

2.2 Chemicals

- Bisphenol A, 99% (e.g. Sigma Aldrich, No. 239658)
- $^{13}C_{12}$-Bisphenol A, 99% (e.g. Cambridge Isotope Laboratories, No. CLM-4325)
- Daidzein, $\geq$ 98% (e.g. Fluka, No. D7802)
- D_3-Daidzein (3′,5′,8-d_3), 97% (e.g. Cambridge Isotope Laboratories, No. DLM-4461)
- Equol, $\geq$ 99.0% (e.g. Fluka, No. 45405)
- Genistein, $\geq$ 98% (e.g. Fluka, No. G6649)
- D_4-Genistein (3′,5′,6,8-d_4), 95% (e.g. Cambridge Isotope Laboratories, No. DLM-4460)
- Acetonitrile, suprasolv (e.g. Merck, No. 100017)
- Ascorbic acid, $\geq$ 99% (e.g. Sigma, No. A5960)
- Glacial acetic acid p.a. (e.g. Merck, No. 100063)
- Ethyl acetate, for gas chromatography (e.g. Merck, No. 110972)
- Ethylene diamine tetraacetic acid (EDTA), anhydrous (e.g. Aldrich, No. 431788)
- n-Hexane, for gas chromatography (e.g. Merck, No. 104371)
- Methanol p.a. (e.g. Merck, No. 106018)
- Sodium acetate trihydrate p.a. (e.g. Merck, No. 106267)
- MTBSTFA with 1% TBDMCS (N-Methyl-N-tert-butyl dimethylsilyl trifluoroacetamide with 1% tert-butyl dimethylchlorosilane) (e.g. Aldrich, No. 375934)
- β-Glucuronidase (*Escherichia coli* K12) (e.g. Roche, No. 03707580001)
- Sulfatase (Abalone entrails, Type VIII) (e.g. Sigma-Aldrich, No. S9754)
- Water, bidistilled
- Nitrogen 5.0
- Helium 4.6

2.3 Solutions

- Hydrolysis buffer
  13.6 g sodium acetate trihydrate, 1.0 g ascorbic acid and 0.1 g EDTA are weighed into a 200 mL glass beaker and dissolved in approx. 70 mL bidist. water. The solution is then adjusted to pH = 5 with glacial acetic acid, transferred to a 100 mL volumetric flask and made up to the mark with bidist. water. The buffer should be colourless and stored at 4°C.
- β-Glucuronidase solution
  The β-glucuronidase (140 U/mL) dissolved in glycerol can be used directly. Isoflavones which may be present should be cleaned out before use for isoflavone analysis. For this purpose, the glucuronidase is passed through a Lichrolut RP18 column preconditioned with methanol and bidist. water.

- Sulfatase solution
  500 U sulfatase are dissolved in 5 mL bidist. water, divided into aliquots of 100 μL each in 0.5 mL Eppendorf tubes and stored at −20°C (100 U/mL) up to use. After preparation this solution is usable for up to one year when stored at −20°C. Once thawed, solutions should not be deep frozen again.
- Enzyme mixture (β-glucuronidase/sulfatase, 2:1)
  200 μL of the β-glucuronidase solution and 100 μL of the sulfatase solution are pipetted into a 0.5 mL Eppendorf tube and mixed thoroughly. The enzyme mixture must be freshly prepared for every analytical series.
- Wash solution (5% methanol in water)
  About 20 mL bidist. water are placed in a 50 mL volumetric flask, then 2.5 mL methanol are added and the flask is made up to the mark with bidist. water.
- Elution solution (ethyl acetate/acetonitrile, 1:1)
  Exactly 25 mL ethyl acetate are pipetted into a 50 mL volumetric flask and the flask is made up to the mark with acetonitrile.

2.4 Internal standards

- Stock solution of the internal standards (IS)
  200 μL d_3-daidzein (c = 60 μg/mL), 120 μL d_4-genistein (c = 100 μg/mL) and 60 μL $^{13}C_{12}$-bisphenol A (c = 100 μg/mL) are pipetted into a 2 mL glass vial and 220 μL acetonitrile are added. This solution contains 20 mg/L d_3-daidzein and 20 mg/L d_4-genistein as well as 10 mg/L $^{13}C_{12}$-bisphenol A.
- Working solution of the internal standards
  200 μL of the IS stock solution are pipetted into a 2 mL vial and mixed with 1800 μL bidist. water. The working solution contains 2 mg/L d_3-genistein and 2 mg/L d_4-daidzein as well as 1 mg/L $^{13}C_{12}$-bisphenol A.

2.5 Calibration standards

- Stock solution genistein (50 mg/L)
  Exactly 5 mg genistein are weighed into a 100 mL volumetric flask and made up to the mark with methanol.
- Stock solution daidzein (c = 50 mg/L)
  Exactly 5 mg daidzein are weighed into a 100 mL volumetric flask and made up to the mark with methanol.
- Stock solution equol (c = 50 mg/L)
  Exactly 1 mg equol is weighed into a 20 mL volumetric flask and made up to the mark with methanol.
- Stock solution bisphenol A (c = 2.2 g/L)
  Exactly 110 mg BPA are placed in a 50 mL volumetric flask and made up to the mark with methanol.

The stability of the stock solutions was not checked. However, no relevant changes in calibrations which were prepared from the same stock solution were found over a period of > 3 months.

Dilutions for calibration

- Bisphenol A predilution (c = 22 mg/L)
  1 mL of the BPA stock solution is placed in a 100 mL volumetric flask and made up to the mark with methanol.
- Working solution I (c = 5 mg/L or 1.1 mg/L)
  1 mL of each of the stock solutions of equol, daidzein and genistein as well as 500 µL of the bisphenol A predilution are pipetted into a 10 mL volumetric flask and made up to the mark with bidist. water. Working solution I contains 5 mg/L each of EQ, DAI and GEN as well as 1.1 mg/L of BPA.
- Working solution II (c = 0.5 mg/L or 0.11 mg/L)
  1 mL of working solution I is pipetted into a 10 mL volumetric flask and made up to the mark with bidist. water. Working solution II contains 0.5 mg/L each of EQ, DAI and GEN as well as 0.11 mg/L of BPA.

Calibration

The calibration standards are prepared in 1.5 mL Eppendorf tubes according to the pipetting scheme in Table 2. For calibration, isoflavone free pooled urine should be used which has been obtained from volunteers who have been on an isoflavone free diet (in particular no soybean products) for at least 2.5 days. As far as possible, the blank value of the prepared calibration series should show no quantifiable analyte signals.

Table 2 Pipetting scheme for the preparation of the calibration standards.

Calibration standard	Spiked analyte level [µg/L]		Pooled urine [µL]	Working soln. I [µL]	Working soln. II [µL]	IS Working soln. [µL]
	EQ, DAI, GEN	BPA				
Blind value	–	–	200	–	–	–
K0	0	0	200	–	–	8
K1	12.5	2.8	200	–	5	8
K2	25.0	5.5	200	–	10	8
K3	50.0	11.0	200	–	20	8
K4	100	22.0	200	4	–	8
K5	250	55.0	200	10	–	8

3 Specimen collection and sample preparation

3.1 Specimen collection

No fixed sampling time is required for the determination of urinary background levels. The urine samples are kept in polyethylene containers at –20°C until preparation.

3.2 Sample preparation

The urine sample is thawed before analysis and equilibrated to room temperature. A 200 µL aliquot of the sample is transferred to a 1.5 mL Eppendorf tube, to which 8 µL of the internal standard working solution is added. Subsequently, the mixture is shaken thoroughly.

Hydrolysis

10 µL of the hydrolysis buffer and 12 µL of the enzyme mixture are added to the sample, which is then incubated overnight (17 h) in a thermostatic shaker at 37°C using the maximum mixing frequency.

Solid phase extraction

The SPE C18 columns are conditioned using 300 µL methanol and 300 µL bidist. water that are subsequently transferred and passed through each column using the vaccum pump. Care must be taken that the SPE columns never completely run dry!

The sample hydrolysed overnight is centrifuged at 16000 g for 1 min. Then the complete sample solution is transferred to the conditioned SPE column using a 300 µL pipette and passed through the SPE column under vacuum. Care must again be taken that the SPE column does not completely run dry. The column is then washed with 300 µL of the wash solution (5% methanol in water) and then dried under vacuum for 10 min. Elution of the analytes is carried out with 300 µL of the elution solution (ethyl acetate / acetonitrile) in a new 1.5 mL Eppendorf tube. For this purpose the elution solution is passed slowly through the SPE column under slight vacuum. Though a formation of two phases in the Eppendorf tube indicates insufficient drying, this has no negative influence on further sample preparation. The sample is then evaporated to dryness using a vacuum centrifuge at 40°C (approx. 1 up to 2 h).

Derivatisation (silylation)

50 µL of the silylation agent (MTBSTFA with 1% TBDMCS) is added to the dried sample, which is then incubated for about 2 min in the ultrasonic bath and briefly vortexed. The sample is then incubated for 30 min at 75°C in the thermostatic shaker (maximum mixing frequency). After cooling to room temperature the sample solution is evaporated to dryness in a stream of nitrogen under gentle heating (40 °C).

After addition of 70 µL n-hexane the sample is incubated in the ultrasonic bath for 1 min, briefly vortexed and then centrifuged at 16000 g for 1 min. The supernatant is then transferred with a pipette into a 2 mL sample vial with a micro-insert and is immediately analysed.

4 Operational parameters

The analysis of the samples prepared according to Section 3 is carried out using a GC-MS unit (e.g. Agilent GC 6890 with Quadrupol MS 5973N and Gerstel MPS2) with following system parameters:

4.1 Operational parameters for gas chromatography

Capillary column:	Material:	Fused Silica
	Stationary phase:	DB-1ms
	Length:	15 m
	Inner diameter:	0.25 mm
	Film thickness:	0.25 µm
Temperature:	Column:	2 min at 160°C; then increase at 30°C/min to 310°C; 5 min at final temperature
	Injector:	0.08 min at 25°C; then increase at 12°C/s to 320°C; hold for 2 min, then increase at 12°C/s to 400°C
	Transfer line:	280°C
Injection:	Liner:	standard liner baffled
	Injection method:	Cooled injection system (CIS 4 plus): 20 µL solvent vent injection, 200 mL/min Helium; solvent vent (in the split vent: 100 mL/min for complete liner heating)
	Injection velocity:	4 µL/s
Carrier gas:	Helium 4.6	
	Constant flow:	2.0 mL/min

4.2 Mass spectrometric parameters

Ionisation type:	Electron impact ionisation (EI)	
Temperatures:	Source:	230°C
	Quadrupol:	150°C
Ionisation energy:	70 eV	

All other parameters must be optimised in accordance with the manufacturer's instructions.

5 Analytical determination

For an additional enrichment of the analytes a cooled injection system is used for gas chromatographic analysis in which 20 µL sample solution are injected and reduced by evaporation. A blank value and a quality control sample as well as a high and a low calibration standard are included for control with each analytical series.

The time profiles of the ion traces listed in Table 3 are recorded in the single ion monitoring (SIM) mode.

The retention times given in Table 3 are intended to be a rough guide only. Users of the method must ensure proper separation performance of the analytical column they use and of the resulting retention behaviour of the analyte.

Table 3 Retention times and recorded masses.

Parameter	Retention time [min]	Quantifier [m/z]	Qualifier [m/z]
Bisphenol A	5.1	441	213, 456
$^{13}C_{12}$-Bisphenol A	5.1	453	225, 468
Equol	6.0	470	234
Daidzein	6.8	425	482
D_3-Daidzein	6.8	428	485
Genistein	7.2	555	–
D_4-Genistein	7.2	558	–

In Figure 3 the chromatogram of a calibration standard is shown which was produced using an injection volume of 20 µL in the solvent vent (cooled injection system). Figure 4 shows a chromatogram of the standard solutions of BPA, daidzein and genistein together with the corresponding internal standards.

6 Calibration

The calibration standards are prepared according to the instructions in Section 3.2 and analysed using GC-MS according to Sections 4 and 5. The calibration graphs are obtained by plotting the ratio of the analyte signal to the internal standard signal as a function of the analyte concentration of the calibration standards. As no isotopically labelled standard is available for equol, $^{13}C_{12}$-BPA is used as an internal standard for both, equol and bisphenol A. $^{13}C_{12}$-BPA showed better results for equol in comparison to the isotopically labelled isoflavones. For routine analysis it is recommended to carry out a complete calibration once per week. Otherwise a blank value, a quality control sample as well as a high and a low calibration standard should be included within each analytical series. The calibration graph is linear between the detection limits and concentrations of 250 µg DAI, 250 µg GEN,

250 µg EQ or 110 µg BPA each per litre urine. Figure 5 shows an example of the calibration graphs for the four analytes.

7 Calculation of the analytical result

To determine the analyte concentration in a sample, the peak area of each analyte is divided by the peak area of the respective internal standard. The ratio thus obtained, is inserted into the corresponding calibration graph (see Section 6). The analyte concentration in µg per litre urine is obtained.

8 Standardisation and quality control

Quality control of the analytical results is carried out as stipulated in the guidelines of the *Bundesärztekammer* (German Medical Association) [51–53]. To obtain reliable results it is recommended to perform double determinations.

To check the precision a spiked urine sample is analysed within each analytical series, which contains a constant and known concentration level of the analytes. As a material for quality control is not commercially available, the control material must be prepared in the laboratory by spiking of isoflavone free pooled urine (see Section 2.5). The quality control material is then divided into aliquots and stored at –20°C. The analyte concentration level of the control material should lie within the relevant concentration range. The nominal value and the tolerance ranges of the quality control material are determined in a pre-analytical period [53].

With the analysis of two calibration standards as well as of one reagent blank value per day of analysis, it is not necessary to include a complete calibration with each analytical series. A new calibration curve should be plotted when the results of quality control indicate systematic deviations of more than 15–20%. In any case, a blank value should be included within each analytical series to detect possible contaminations.

9 Evaluation of the method

9.1 Precision

To determine the within day precision, spiked urine samples with low and high analyte levels were analysed four times in a row. The precision data thus obtained are listed in Table 4.

To determine the precision from day to day, urine samples spiked with standard solutions were prepared and analysed on six different days. The results are listed in Table 5.

Table 4 Within day precision (n = 4).

Parameter	Spiked analyte level [µg/L]		Standard deviation (rel.) [%]		Prognostic range [%]	
	Urine 1	Urine 2	Urine 1	Urine 2	Urine 1	Urine 2
Genistein	10	125	17	3	54	10
Daidzein	10	125	5	7	16	22
Equol	10	125	6	5	19	16
Bisphenol A	2.2*	27.5	7	5	22	16

* Although this concentration is below the given detection limit, it can nevertheless be determined with a certain reliability. However, more extensive deviations were occasionally found in the low concentration range (see Section 9.5).

Table 5 Day to day precision (n = 6).

Parameter	Spiked analyte level [µg/L]	Standard deviation (rel.) [%]	Prognostic range [%]
Genistein	75	18	46
Daidzein	75	9	23
Equol	75	15	39
Bisphenol A	16.5	10	26

9.2 Accuracy

Recovery experiments with spiked urine samples were carried out to determine the accuracy of the method. For this purpose pooled urine was spiked with defined analyte levels, processed four times and analysed. The determinations were carried out both within day as well as from day to day. The mean relative recovery rates thus obtained are shown in Table 6.

Table 6 Relative recovery rates within day and from day to day (n = 4).

Parameter	Spiked analyte level [µg/L]	Mean relative recovery [%] (n = 4)	
		Within day	From day to day
Genistein	10	119	–
	75	–	126
	125	114	105
Daidzein	10	108	–
	75	–	127
	125	109	104
Equol	10	104	–
	75	–	90
	125	89	112
Bisphenol A	16.5	–	119
	27.5	94	90

9.3 Detection limits and quantitation limits

The detection and quantitation limits were calculated in accordance with DIN method 32645 [German Industrial Standard, *Deutsche Industrie Norm*] (calibration graph method). For this purpose, three independent calibration graphs with 10 calibration points per analyte were performed. The results are listed in Table 7.

Table 7   Detection and quantitation limits, calculated according to DIN 32645.

Parameter	Detection limit [µg/L]	Quantitation limit [µg/L]
Genistein	5	18
Daidzein	4	9
Equol	4	11
Bisphenol A	3	7

9.4 Validation data of the alternative method

As an alternative for direct splitless injection of 20 µL of the prepared sample solution (see Section 3.2) the method can also be carried out without the use of a cooled injection system. For this only 2 µL of the sample solution are injected directly (splitless) into the GC-MS system. This procedure was applied as part of the successful examination of the method. The validation data thus obtained are described in Table 8. In spite of the 10-fold lower injection volume the detection limits using the alternative injection method are only 2–4 times higher.

Table 8   Validation data when using an injection volume of 2 µL.

Parameter	Spiked analyte level [µg/L]	BPA	Equol	Daidzein	Genistein
Detection limit [µg/L]		5	15	11	9
Quantitation limit [µg/L]		9	29	26	20
Within day precision [%, n = 6]	50	8	12	8	11
	125	5	9	5	7
Day to day precision [%, n = 4]	50	7	13	9	12
	125	8	9	8	9
Rel. recovery rate (Accuracy) [%, n = 4]	30	109 ± 10	102 ± 6	110 ± 5	115 ± 7
	50	101 ± 4	102 ± 4	100 ± 5	99 ± 5
	125	100 ± 4	100 ± 2	99 ± 4	98 ± 4
Absolute recovery after extraction [%, n = 6]		81 ± 2	88 ± 4	85 ± 5	90 ± 2

9.5 Sources of error

BPA is ubiquitous and may also be found in house dust. During sample preparation special cleanliness must be ensured. Insufficiently washed containers may otherwise produce a considerable blank value for BPA. The bidistilled water used may also contain BPA. Therefore the water used should previously be cleaned by passing it through a Lichrolut RP 18 cartridge preconditioned with methanol. A reagent blank value is therefore to be included in every analytical series to monitor the blank value.

The fragmentation pattern of the analytes in the mass spectrometer is greatly influenced by the pH value of the sample. Therefore, as part of the sample preparation, the urine samples are buffered to avoid interferences. The analysis of highly concentrated urine samples, especially urine samples from volunteers whose diet has a high soy content (e.g. Asian cuisine) can lead to carry-over contaminations when using the cooled injection system. In this case the heating of the liner should be extended or modified or alternatively 2 µL of the sample should be injected (see also Section 10).

In order to minimise the memory effect, a regular injection of pure n-hexane between the samples is recommended, especially after injection of highly concentrated samples. If the linear range of the method is exceeded, the samples should be diluted and processed again.

10 Discussion of the method

The present method was developed based on the studies by Liggins et al., Setchell et al. and Moors et al. [54–56] and allows a sensitive determination of substances with estrogenic activity, like genistein, daidzein, equol and bisphenol A. As the listed substances are mainly eliminated with the urine in the form of conjugates, urinary determination is carried out after hydrolysis, in which glucuronides and sulfates are cleaved. In order to guarantee a sensitive detection using GC-MS, the determination of the analytes is carried out after derivatisation of the free hydroxyl groups. In Figure 7 the applied silylation reaction is described using GEN as example.

The importance of the method lies mainly in the field of environmental medicine, as the phytoestrogens DAI and GEN are almost exclusively taken in via food. The large-scale chemical product bisphenol A, to which an estrogenic activity is also attributed, is considered to be relevant for occupational medicine as well.

The simultaneous detection of these four important hormonally active substances makes the analytical method applicable and useful for both, occupational and environmental medicine.

In the present method, sensitivity is increased when using a cooled injection system, whereby 20 µL of the sample solution is injected and reduced by evaporation with the aid of a temperature program in the injector block. Alternatively, it is also possible to apply the method by injecting 2 µL of the prepared sample directly into the GC-MS unit. The detection limits are 2 to 4 times higher when using this alternative method. The linear working range of the method is, however, limited by the peak detector capacity; in the case of the Agilent MSD 5973N mass spectrometer used here, this is limited to 130 million area units. With a volume of 20 µL used in the cooled injection mode this limit was approximately 300 to 400 µg analyte per litre urine. Urine samples with such high isoflavone concentrations are, for example, to be expected in persons with a high soy intake. However, with a sample dilution, it is also possible to detect the mentioned analytes in highly concentrated samples.

Instruments used:

- Gas chromatograph (GC 6890, Agilent, USA) with autosampler (MPS2, Gerstel, Germany) and with a Quadrupole-mass spectrometer (5973N, Agilent, USA).

11 References

1 National Research Council: Hormonally Active Agents in the Environment. National Academy Press, Washington DC (1999).
2 G.H. Degen and H.M. Bolt: Endocrine disruptors: Update on xenoestrogens. Int. Arch. Occup. Environ. Health 73, 433–441 (2000).
3 S. Fritsche and H. Steinhart: Occurrence of hormonally active compounds in food: a review. Eur. Food Res. Technol. 209, 153–179 (1999).
4 G.H. Degen: Endokrine Disruptoren in Lebensmitteln. Bundesgesundheitsblatt 9, 848-856 (2004).
5 K. Reinli and G. Block: Phytoestrogen content of foods – a compendium of literature values. Nutr. Cancer 26, 123–148 (1996).
6 W.H. Tolleson, D.R. Doerge, M.I. Churchwell, M.M. Marques, D.W. Roberts: Metabolism of biochanin A and formononetin by human liver microsomes in vitro. J. Agric. Food Chem. 50, 4783–4790 (2002).
7 H. Hur and F. Rafii: Biotransformation of the isoflavonoids biochanin A, formononetin, and glycitein by Eubacterium limosum. FEMS Microbiol. Lett. 192, 21–25 (2000).
8 H. Saloniemi, K. Wähälä, P. Nykänen-Kurki, K. Kallela, I. Saastamoinen: Phytoestrogen content and estrogenic effect of legume fodder. Proc. Soc. Exp. Biol. Med. 208, 13–17 (1995).
9 W. Mazur: Phytoestrogen content in foods. Baillieres Clin. Endocrinol. Metab. 12, 729–742 (1998).
10 A.A. Franke, L.J. Custer, C.M. Cerna, K.K. Narala: Quantitation of phytoestrogens in legumes by HPLC. J. Agric. Food Chem. 42, 1905–1913 (1994).
11 J. Liggins, L.J. Bluck, S. Runswick, C. Atkinson, W.A. Coward, S.A. Bingham: Daidzein and genistein content of fruits and nuts. J. Nutr. Biochem. 11, 326–331 (2000).
12 O. Lapcik, M. Hill, R. Hampl, K. Wähälä, H. Adlercreutz: Identification of isoflavonoids in beer. Steroids 63, 14–20 (1998).

13 U.S. Department of Agriculture, Agricultural Research Service: USDA-Iowa State University Database on the Isoflavone Content of Foods. Release 1.3 (2002). URL: http://www.nal.usda.gov/fnic/foodcomp/Data/isoflav/isoflav.html visited on 06.06.2011.
14 L.K. Boker, Y.T. Van der Schouw, M.J. De Kleijn, P.F. Jacques, D.E. Grobbee, P.H. Peeters: Intake of dietary phytoestrogens by Dutch women. J. Nutr. 132, 1319–1328 (2002).
15 M.J. De Kleijn, Y.T. Van der Schouw, P.W. Wilson, H. Adlercreutz, W. Mazur, D.E. Grobbee, P.F. Jaques: Intake of dietary phytoestrogens is low in postmenopausal women in the United States: the Framingham study (1–4). J. Nutr. 131, 1826–1832 (2001).
16 P.L. Horn-Ross, S. Barnes, M. Lee, L. Coward, J.E. Mandel, J. Koo, E.M. John, M. Smith: Assessing phytoestrogen exposure in epidemiologic studies: development of a database (United States). Cancer Causes Control 11, 289–298 (2000).
17 Y. Arai, M. Uehara, Y. Sato, M. Kimira, A. Eboshida, H. Adlercreutz, S. Watanabe: Comparison of isoflavones among dietary intake, plasma concentration and urinary excretion for accurate estimation of phytoestrogen intake. J. Epidemiol. 10, 127–135 (2000).
18 T. Usui: Pharmaceutical prospects on phytoestrogens. Endocr. J. 53, 7–20 (2006).
19 E. Bowey, H. Adlercreutz, I. Rowland: Metabolism of isoflavones and lignans by the gut microflora: a study in germfree and human flora associated rats. Food Chem. Toxicol. 41, 631–636 (2003).
20 L.N. Vandenberg, M.V. Maffini, C. Sonnenschein, B.S. Rubin, A.M. Soto: Bisphenol-A and the great divide: a review of controversies in the field of endocrine disruption. Endocrine Reviews 30, 75–95 (2009).
21 H. Greim (ed.): Bisphenol A. MAK Value Documentations. Volume 13, VCH-Verlag, Weinheim (1999).
22 H. Greim and H. Drexler (eds.): Bisphenol A. Biologische Arbeitsstoff-Toleranz-Werte (BAT-Werte) und Expositionsäquivalente für krebserzeugende Arbeitsstoffe (EKA) und Biologische Leitwerte (BLW). 14th issue, Wiley-VCH-Verlag, Weinheim (2007).
23 A. Goodson, W. Summerfield, I. Cooper: Survey of bisphenol A and bisphenol F in canned foods. Food Addit. Contam. 19, 796–802 (2002).
24 B.L. Tan and A.M. Mustafa: Leaching of bisphenol A from new and old babies' bottles, and new babies' feeding teats. Asia Pac. J. Public Health 15, 118-123 (2003).
25 Scientific Committee on Food: Opinion of the Scientific Committee on Food on Bisphenol A (expressed on 17 April 2002). European Commission, URL: http://ec.europa.eu/food/fs/sc/scf/out128_en.pdf, visited on 06.06.2011.
26 G.H. Degen, H. Foth, R. Kahl, H. Kappus, H.G. Neumann, F. Oesch, R. Schulte-Hermann: Hormonell aktive Substanzen in der Umwelt: Xenoöstrogene. Stellungnahme der Beratungskommission der Sektion Toxikologie der DGPT. Umweltmed. Forsch. Praxis 4, 367–374 (1999).
27 G.H. Degen, P. Janning, J. Wittsiepe, A. Upmeier, H.M. Bolt: Integration of mechanistic data in the toxicological evaluation of endocrine modulators. Toxicol. Lett. 127, 225–237 (2002).
28 G. Eisenbrand, B. Mußler, G.H. Degen: Kombinationswirkungen hormonartig wirkender Chemikalien. In: H.M. Bolt, B. Griefahn, H. Heuer, W. Laurig (Hrsg): Arbeitsphysiologie heute. Band 2. IfADo, Dortmund, S. 27–39 (2000).
29 P. Diel, T. Schulz, K. Smolnikar, E. Strunck, G. Vollmer, H. Michna: Ability of xeno- and phytoestrogens to modulate expression of estrogen-sensitive genes in rat uterus: estrogenicity profiles and uterotrophic activity. J. Steroid Biochem. Mol. Biol. 73, 1–10 (2000).
30 P. Diel, S. Schmidt, G. Vollmer, P. Janning, A. Upmeier, H. Michna, H.M. Bolt, G.H. Degen: Comparative responses of three rat strains (DA/Han, Sprague-Dawley and Wistar) to treatment with environmental estrogens. Arch. Toxicol. 78, 183–193 (2004).

31 S. Schmidt, G.H. Degen, J. Seibel, T. Hertrampf, G. Vollmer, P. Diel: Hormonal activity of combinations of genistein, bisphenol A and 17ß-estradiol in the female Wistar rat. Arch. Toxicol. 80, 839–845 (2006).
32 P. Janning, U.S. Schuhmacher, A. Upmeier, P. Diel, H. Michna, G.H. Degen, H.M. Bolt: Toxicokinetics of the phytoestrogen daidzein in female DA/Han rats. Arch. Toxicol. 74, 421–430 (2000).
33 T. Bayer, T. Colnot, W. Dekant: Disposition and biotransformation of the estrogenic isoflavone daidzein in rats. Toxicol. Sci. 62, 205–211 (2001).
34 G.H. Degen, P. Janning, P. Diel, H. Michna, H.M. Bolt: Transplacental transfer of the phytoestrogen daidzein in DA/Han rats. Arch. Toxicol. 76, 23–29 (2002).
35 R.W. Snyder, S.C. Maness, K.W. Gaido, F. Welsch, S.C. Sumner, T.R. Fennell: Metabolism and disposition of bisphenol A in female rats. Toxicol. Appl. Pharmacol. 168, 225–234 (2000).
36 S. Moors, P. Diel, G.H. Degen: Toxicokinetics of bisphenol A in pregnant DA/Han rats after single i.v. application. Arch. Toxicol. 80, 647–655 (2006).
37 J.W. Lampe: Isoflavonoid and lignan phytoestrogens as dietary biomarkers. J. Nutr. 133, 956S–964S (2003).
38 W. Völkel, T. Colnot, G.A. Csanády, J.G. Filser, W. Dekant: Metabolism and kinetics of bisphenol A in humans at low doses following oral administration. Chem. Res. Toxicol. 15, 1281–1287 (2002).
39 L. Valentín-Blasini, M.A. Sadowski, D. Walden, L. Caltabiano, L.L. Needham, D.B. Barr: Urinary phytoestrogen concentrations in the U.S. population (1999–2000). J. Expo. Anal. Environ. Epidemiol. 15, 509–523 (2005).
40 Centers for Disease Control and Prevention (CDC): Third National Report on Human Exposure to Environmental Chemicals. CDC, Atlanta (GA), S. 285–306 (2005).
41 K.D. Setchell, N.M. Brown, L. Zimmer-Nechemias, W.T. Brashear, B.E. Wolfe, A.S. Kirschner, J.E. Heubi: Evidence for lack of absorption of soy isoflavone glycosides in humans, supporting the crucial role of intestinal metabolism for bioavailability. Am. J. Clin. Nutr. 76, 447–453 (2002).
42 M.S. Faughnan, A. Hawdon, E. Ah-Singh, J. Brown, D.J. Millward, A. Cassidy: Urinary isoflavone kinetics: the effect of age, gender, food matrix and chemical composition. Br. J. Nutr. 91, 567–574 (2004).
43 A.J. Day, M.S. DuPont, S. Ridley, M. Rhodes, M.J. Rhodes, M.R. Morgan, G. Williamson: Deglycosylation of flavonoid and isoflavonoid glycosides by human small intestine and liver beta-glucosidase activity. FEBS Lett. 436, 71–75 (1998).
44 J.B. Matthews, K. Twomey, T.R. Zacharewski: In vitro and in vivo interactions of bisphenol A and its metabolite, bisphenol A glucuronide, with estrogen receptor alpha and beta. Chem. Res. Toxicol. 14, 149–157 (2001).
45 E. Federici, A. Touché, S. Choquart, O. Avanti, L. Fay, E. Offord, D. Courtois: High isoflavone content and estrogenic activity of 25 year-old Glycine max tissue cultures. Phytochemistry 64, 717–724 (2003).
46 Y. Zhang, T.T. Song, J.E. Cunnick, P.A. Murphy, S. Hendrich: Daidzein and genistein glucuronides in vitro are weakly estrogenic and activate human natural killer cells at nutritionally relevant concentrations. J. Nutr. 129, 399–405 (1999).
47 S.O. Mueller, S. Simon, K. Chae, M. Metzler, K.S. Korach: Phytoestrogens and their human metabolites show distinct agonistic and antagonistic properties on estrogen receptor alpha (ERalpha) and ERbeta in human cells. Toxicol. Sci. 80, 14–25 (2004).
48 N. Sathyamoorthy and T.T. Wang: Differential effects of dietary phytoestrogens daidzein and equol on human breast cancer MCF-7 cells. Eur. J. Cancer 33, 2384–2389 (1997).

49 I.R. Rowland, H. Wiseman, T.A. Sanders, H. Adlercreutz, E.A. Bowey: Interindividual variation in metabolism of soy isoflavones and lignans: influence of habitual diet on equol production by the gut microflora. Nutr. Cancer 36, 27–32 (2000).

50 P.B. Grace, J.I. Taylor, N.P. Botting, T. Fryatt, M.F. Oldfield, S.A. Bingham: Quantification of isoflavones and lignans in urine using gas chromatography/mass spectrometry, Anal. Biochem. 315, 114–121 (2003).

51 Bundesärztekammer: Qualitätssicherung der quantitativen Bestimmungen im Laboratorium. Neue Richtlinien der Bundesärztekammer. Dt. Ärztebl. 85, A699-A712 (1988).

52 Bundesärztekammer: Ergänzung der "Richtlinien der Bundesärztekammer zur Qualitätssicherung in medizinischen Laboratorien". Dt. Ärztebl. 91, C159–C161 (1994).

53 J. Angerer, T. Göen, G. Lehnert: Mindestanforderungen an die Qualität von umweltmedizinisch-toxikologischen Analysen. Umweltmed. Forsch. Prax. 3, 307–312 (1998).

54 J. Liggins, L. Bluck, W.A. Coward, S.A. Bingham: A simple method for the extraction and quantification of daidzein and genistein in food using gas chromatography mass spectrometry. Biochem. Soc. Trans. 26, S87 (1998).

55 K.D. Setchell, N.M. Brown, P. Desai, L. Zimmer-Nechemias, B.E. Wolfe, W.T. Brashear, A.S. Kirschner, A. Cassidy, J.E. Heubi: Bioavailability of pure isoflavones in healthy humans and analysis of commercial soy isoflavone supplements. J. Nutr. 131, 1362S–1375S (2001).

56 S. Moors, M. Blaszkewicz, H.M. Bolt, G.H. Degen: Simultaneous determination of daidzein, equol, genistein and bisphenol A in human urine by a fast and simple method using SPE and GC-MS. Mol. Nutr. Food Res. 51, 787–798 (2007).

Authors: *S. Moors, M. Blaszkewicz*
Examiner: *D. Barr*

12 Appendix

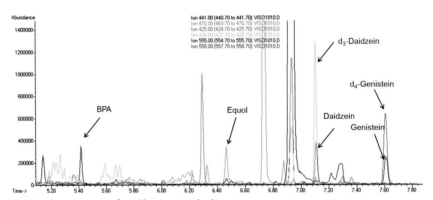

Fig. 3 Chromatogram of a calibration standard in urine.

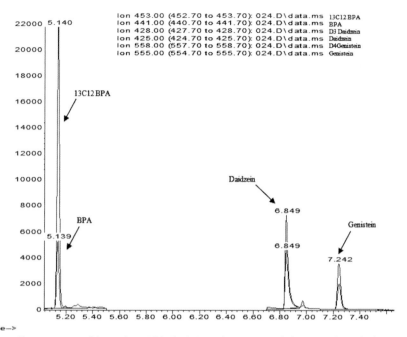

Fig. 4 Chromatogram of the analytes BPA, daidzein and genistein with the internal standards.

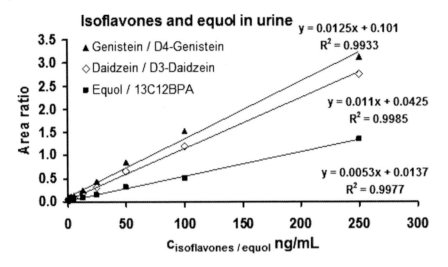
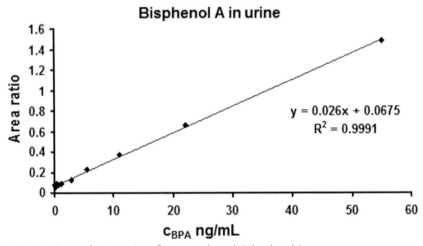

Fig. 5   Calibration functions: A: isoflavones and equol, B: bisphenol A.

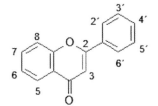

Flavanones
Naringenin: 4',5,7-Trihydroxyflavanone
Sakuranetin: 4',5-Dihydroxy-7-methoxyflavanone

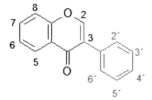

Flavones
Primuletin: 5-Hydroxyflavone
Chrysin: 5,7-Hydroxyflavone
Apigenin: 4',5,7-Hydroxyflavone
Kaempferol: 3,4',5,7-Tetrahydroxyflavone

Isoflavones
Daidzein: 4',7-Dihydroxyisoflavone
Formononetin: 7-Hydroxy-4'-methoxyisoflavone
Genistein: 4',5,7-Trihydroxyisoflavone
Biochanin A: 5,7-Dihydroxy-4'-methoxyisoflavone

Fig. 6 Basic structure of the flavanones, flavones and isoflavones.

Fig. 7 Derivatisation reaction using genistein as example.

N-(3-Chloro-2-hydroxypropyl)-valine in blood as haemoglobin adduct of epichlorohydrin

Matrix:	Blood
Hazardous substance:	Epichlorohydrin
Analytical principle:	Capillary gas chromatography with tandem mass spectrometry (GC-MS/MS)
Completed in:	May 2009

Overview of the parameters that can be determined with this method and the corresponding chemical hazardous substances:

Hazardous substance	CAS	Parameter	CAS
Epichlorohydrin	106-89-8	N-(3-Chloro-2-hydroxypropyl)-valine	–

Summary

With this analytical method the primary haemoglobin adduct of epichlorohydrin, N-(3-chloro-2-hydroxypropyl)-valine (CHPV) can be determined quantitatively using gas chromatography-tandem mass spectrometry (GC-MS/MS).

For this purpose, globin is isolated from a blood sample by precipitation, dried and dissolved in formamide. After addition of pentafluorophenyl isothiocyanate, sodium hydroxide and the internal standard (d_5-CHPV-labelled globin), selective cleavage is carried out via a modified Edman degradation including conversion of the adduct to a thiohydantoin. The derivative is extracted with diethyl ether, dried in a stream of nitrogen and resuspended in toluene. After washing using sodium carbonate solution and water the sample is evaporated in a stream of nitrogen. The hydroxypropyl function of the adduct is acetylated by addition of acetic anhydride/triethylamine in acetonitrile. The sample is then evaporated to dryness in a stream of nitrogen, dissolved in n-hexane and extracted with a mixture of methanol and water. The hexane phase is evaporated to dryness in a stream of nitrogen and the residue is dissolved in toluene. Quantitative analysis of the thiohydantoin derivative is carried out using GC-MS/MS in the selected reaction monitoring (SRM) mode after negative chemical ionisation (NCI).

For external calibration the calibration standards of an adduct-modified dipeptide, N-(3-chloro-2-hydroxypropyl)-valine leucine anilide are used.

N-(3-Chloro-2-hydroxypropyl)-valine (CHPV)

Within day precision: Standard deviation (rel.) s_w = 12.4% or 9.8%
Prognostic range u = 27.6% or 21.8%
at a spiked concentration of 25 or 100 pmol CHPV per gram globin and where n = 10 determinations

Day to day precision: Standard deviation (rel.) s_w = 15.0%
Prognostic range u = 33.4%
based on the variance of the calibration graph slope and where n = 10 determinations

Accuracy: Recovery rate r = 99.8 ± 12.5%
at a nominal concentration of 100 pmol CHPV per gram globin and where n = 10 determinations

Detection limit: 10 pmol CHPV per gram globin
Quantitation limit: 25 pmol CHPV per gram globin

1-Chloro-2,3-epoxypropane (epichlorohydrin)

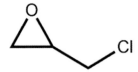

1-Chloro-2,3-epoxypropane (epichlorohydrin, CAS-No. 106-89-8) is a chloromethyl-substituted epoxide with an annual production volume of around 50,000 tons. Epichlorohydrin is mainly used for the production of epoxy resins, glycerol and insecticides and as cross-linking agent in the paper industry [1]. At room temperature, epichlorohydrin is a colourless liquid with a pungent odour, a boiling point of 116°C and a vapour pressure of about 17 hPa. It is poorly soluble in water [2]. In a neutral aqueous solution epichlorohydrin hydrolyses to 3-chloropropane-1,2-diol with a half-life of 5 to 8 days. This takes place more rapidly in acidic or alkaline solutions [1].

In the 2003 MAK documentation, the *Deutsche Forschungsgemeinschaft* (DFG) summarised the state of knowledge on the toxicity of epichlorohydrin. The DFG has classified it in Category 2 for carcinogenic substances [2]. Analogously, the IARC (International Agency for Research on Cancer) has categorised epichlorohydrin as probably carcinogenic to humans (Group 2A) [3]. Furthermore, epichlorohydrin has been designated with "Sh" (skin sensitisation hazard) and with an "H" (risk of skin absorption) and it is also classified as a Category 3B germ cell mutagen due to its genotoxicity [2].

Knowledge about the metabolism of epichlorohydrin is mainly derived from animal studies by Weigel et al. [4] and Gingell et al. [5]. According to these studies, about 90% of an inhaled or orally administered dose is absorbed and distributed throughout all body compartments within 2 to 4 hours. Metabolism primarily takes place via reaction of the two electrophilic centres of epichlorohydrin, the C3 in the epoxide ring and the chlorine atom at the C1. Accordingly, conjugation with glutathione leads to the excretion of N-acetyl-S-(3-chloro-2-hydroxypropyl)-L-cysteine, S-(2,3-dihydroxypropyl)-L-cysteine and N-acetyl-S-(2,3-dihydroxypropyl)-L-cysteine as main metabolites. In addition, 3-chloropropane-1,2-diol is formed by epoxide hydrolysis.

In vivo, epichlorohydrin is able to react with macromolecules. Binding to haemoglobin and to lymphocyte DNA has already been described a number of times in the literature and has been considered for biomonitoring [6–13]. In principle, two types of adducts can be distinguished and used for analysis (see Figure 1). These are on the one hand the primary adducts with a 3-chloro-2-hydroxypropyl structure formed by epoxide ring opening and on the other hand the secondary adducts with a 2,3-dihydroxypropyl structure. These secondary adducts are formed either by hydrolytic cleavage of the chlorine-carbon bond after conjugation to glutathione or by direct conjugation of 3-chloro-1,2-propanediol to glutathione under HCl elimination.

Hindsø Landin et al. [9] have been able to show that the concentration of 3-chloro-2-hydroxypropyl-cysteine adducts in rat globin decreases more rapidly ($t_{1/2}$ approx. 4 days) than expected from the lifespan of rat erythrocytes (approx. 30 days). In contrast, the concentration of 2,3-dihydroxypropyl-valine adducts in rats' blood increased after exposure. In 1996, the working group of Hindsø Landin was not successful in analysing 3-chloro-2-hydroxypropyl-valine [14]. From these observations they assumed that the chlorinated primary adduct is not very stable neither *in vivo* nor *in vitro* and is therefore not suitable for biomonitoring [9]. However, more recent studies mentioned above showed that a determination of the 3-chloro-2-hydroxypropyl adduct is possible. At present, human *in vivo* studies are rare. Primarily, occupational medical studies involving employees exposed to epichlorohydrin and investigations following single accidental exposure were carried out.

The DNA adduct N7-(3-chloro-2-hydroxypropyl)-guanine was analysed in a study by Plna et al. [12] and was not detected in 13 samples of controls but in 7 of 16 samples of volunteers potentially exposed to epichlorohydrin.

In 1997, Hindsø Landin et al. determined N-(2,3-dihydroxypropyl)-valine (DHPV) using a GC-MS/MS method (detection limit: 2 pmol/g globin) in 15 German employees exposed to epichlorohydrin (determined DHPV levels were: eight non-smokers: 7.3 ± 2.7 pmol/g globin; seven smokers: 21.1 ± 17.1 pmol/g globin) as well as in eleven controls from the same company (DHPV concentration: three non-smokers: 6.8 ± 3.2 pmol/g globin; nine smokers: 13.1 ± 12.4 pmol/g globin) [8]. The mean external exposure of the workers exposed to epichlorohydrin during a 12-hour work shift ranged between 0.11 and 0.23 ppm (former Technical Guidance Concentration: 3 ppm). The differences between non-smokers and smokers were significant, whereas no differences were found between exposed and control persons. Thus, Hindsø Landin et al. [8] concluded that the dihydroxypropyl-valine adduct was not substance-specific but could also have been formed from glycidol found in tobacco smoke. Accordingly, tobacco smoke-related adduct levels may interfere with adduct levels resulting from workplace exposure.

Furthermore, this working group analysed six non-smokers and four smokers of a Swedish control collective [14]. With an adduct level of 2.1 ± 1.1 pmol/g globin (non-smokers) or 9.5 ± 2.2 pmol/g globin (smokers), the DHPV level of the Swedish volunteers was significantly lower than that of the controls in the German study.

In a study by Wollin et al. [15] with accidentally exposed individuals it was not possible to detect DHPV neither in smokers nor in individuals potentially exposed to epichlorohydrin (GC-MS, detection limit 10 pmol/g globin).

In a collective of 323 emergency helpers potentially exposed to released epichlorohydrin in an accident involving the transport of hazardous materials, Bader et al. [13] were able to detect the primary adduct N-(3-chloro-2-hydroxypropyl)-valine (CHPV) in a total of 22 samples. The level and duration of each exposure were unknown. Furthermore, CHPV was also found *in vitro* after incubation of human erythrocytes with epichlorohydrin. This implies that CHPV can be used as a biomarker for exposure to epichlorohydrin.

Authors: *M. Bader, W. Rosenberger, D. Tsikas, F.M. Gutzki*
Examiners: *Th. Göen, E. Eckert*

N-(3-Chloro-2-hydroxypropyl)-valine in blood as haemoglobin adduct of epichlorohydrin

Matrix:	Blood
Hazardous substance:	Epichlorohydrin
Analytical principle:	Capillary gas chromatography with tandem mass spectrometry (GC-MS/MS)
Completed in:	May 2009

Contents

1 General principles
2 Equipment, chemicals and solutions
2.1 Equipment
2.2 Chemicals
2.3 Solutions
2.4 Internal standard
2.5 Calibration standards
3 Specimen collection and sample preparation
3.1 Preparation of the erythrocyte lysate
3.2 Isolation of the globin
3.3 Derivatisation of the globin
4 Operational parameters
4.1 Operational parameters for gas chromatography
4.2 Operational parameters for tandem mass spectrometry
5 Analytical determination
6 Calibration
7 Calculation of the analytical result
8 Standardisation and quality control
9 Evaluation of the method
9.1 Precision
9.2 Accuracy

9.3   Detection limit and quantitation limit
9.4   Sources of error
10   Discussion of the method
11   References
12   Appendix

1 General principles

With the present analytical method the major haemoglobin adduct of epichlorohydrin, N-(3-chloro-2-hydroxypropyl)-valine can be determined quantitatively based on the so-called modified Edman degradation using gas chromatography-tandem mass spectrometry (GC-MS/MS). For this purpose, globin is initially isolated from a whole blood sample by sequential precipitation, dried and then dissolved in formamide. After the addition of pentafluorophenyl isothiocyanate, sodium hydroxide and an internal standard (d_5-CHPV-labelled globin), selective cleavage and conversion of the N-alkylated terminal amino acid into a thiohydantoin derivative is carried out (Figure 2). The derivative is extracted with diethyl ether, dried in a stream of nitrogen and resuspended in toluene. After washing steps using sodium carbonate solution and water, the sample is evaporated to dryness in a stream of nitrogen. The free hydroxyl function of the adduct is acetylated by the addition of acetic anhydride/triethylamine in acetonitrile. The sample is evaporated to dryness in a stream of nitrogen, dissolved in n-hexane and washed with methanol/water (60:40, v/v). The hexane phase is evaporated to dryness in a stream of nitrogen and the residue is dissolved in toluene. The thiohydantoin derivative is quantitatively analysed using GC-MS/MS in the selected reaction monitoring mode (SRM) after negative chemical ionisation (NCI). Calibration is carried out with a dipeptide standard containing N-terminal N-(3-chloro-2-hydroxypropyl)-valine. A globin labelled with pentadeuterated epichlorohydrin (d_5-epichlorohydrin) is used as internal standard.

2 Equipment, chemicals and solutions

2.1 Equipment

- GC-MS/MS system with PTV injector, autosampler and negative chemical ionisation (NCI) unit (e.g. ThermoElectron (Dreieich, Germany)) with TSQ 7000 mass spectrometer)
- Gas chromatographic column: length: 30 m; inner diameter: 0.25 mm; stationary phase: 100% polydimethylsiloxane; film thickness: 0.25 µm (e.g. Optima-1-MS, Macherey-Nagel, No. 726205.30)

- 10 mL, 50 mL, 100 mL, 500 mL, 1000 mL Volumetric flasks
- 13 mL and 20 mL Screw cap glass tubes (e.g. Schott)
- Laboratory shaker (e.g. Vibrax, Ika)
- Horizontal shaker (e.g. Janke & Kunkel)
- Laboratory centrifuge (e.g. Megafuge, Heraeus)
- Sample evaporation unit (e.g. ReactiVap, Pierce)
- 1.8 mL Glass vials (e.g. Macherey-Nagel)
- 200 µL Microvials (e.g. Macherey-Nagel)
- Water bath with shaking device (e.g. Thermo Haake)
- Microlitre pipettes, with adjustable volumes between 2 and 20 µL, 10 and 100 µL as well as between 100 and 1000 µL (e.g. Eppendorf)
- Analytical balance (e.g. Mettler Toledo)
- Vacuum desiccator (e.g. Schott)
- Ultrasonic bath (e.g. Bandelin electronic)
- EDTA monovettes (e.g. Sarstedt)
- 250 mL Glass beaker

2.2 Chemicals

- 2-Propanol p.a. (e.g. Merck, No. 109634)
- Ethyl acetate p.a. (e.g. Merck, No. 109623)
- n-Hexane p.a. (e.g. Merck, No. 107023)
- Sodium chloride p.a. (e.g. Merck, No. 106404)
- Sodium carbonate, anhydrous p.a. (e.g. Merck, No. 106392)
- 37% Hydrochloric acid (e.g. Merck, No. 100317)
- Sodium hydroxide pellets p.a. (e.g. Merck, No. 106498)
- Highly purified water or bidist. water
- Diethyl ether p.a. (e.g. Merck, No. 100921)
- Formamide ultrapure (e.g. Amersham Life Science, No. US75828)
- Ethanol p.a. (e.g. Merck, No. 100983)
- N-(3-Chloro-2-hydroxypropyl)-valine leucine anilide (e.g. Bachem, No. G-4340)
- Acetic anhydride p.a. (e.g. Merck, No. 100042)
- Acetonitrile ultrapure (e.g. Fluka, No. 00700)
- Triethylamine p.a. (e.g. Merck, No. 808352)
- Methanol p.a. (e.g. Merck, No. 106011)
- d_5-Epichlorohydrin, 98% (e.g. Cambridge Isotope Laboratories, No. DLM-1008-1)
- Pentafluorophenyl isothiocyanate for GC, $\geq$ 97% (e.g. Fluka, No. 76755)
- Toluene p.a. (e.g. Fluka, No. 34938)
- Nitrogen 5.0 (e.g. Linde)
- Helium 5.0 (e.g. Linde)
- Methane 5.5 (e.g. Linde)
- Argon 5.0 (e.g. Linde)

2.3 Solutions

- 50 mM Hydrochloric acid in 2-propanol:
  4.1 mL 37% hydrochloric acid are pipetted into a 1000 mL volumetric flask containing about 500 mL 2-propanol and made up to the mark with 2-propanol.
- 0.9% Saline solution:
  9 g sodium chloride are weighed into a 1000 mL volumetric flask, dissolved in bidist. water and made up to the mark with bidist. water.
- 1 M Sodium hydroxide solution:
  4 g sodium hydroxide pellets are weighed into a 250 mL glass beaker and dissolved in bidist. water. The solution is transferred to a 100 mL volumetric flask and made up to the mark with bidist. water.
- 0.1 M Sodium carbonate solution:
  1.06 g sodium carbonate are weighed into a 100 mL volumetric flask, dissolved in bidist. water and made up to the mark with bidist. water.
- Wash solution:
  60 mL methanol are transferred to a 100 mL volumetric flask and made up to the mark with bidist. water.

The solutions can be stored in the refrigerator at 4°C for at least one month.

Acetylation reagent:

- 12.5% (v/v) Acetic anhydride:
  1.25 mL acetic anhydride are placed in a 10 mL volumetric flask and made up to the mark with acetonitrile.
- 12.5% (v/v) Triethylamine:
  1.25 mL triethylamine are placed in a 10 mL volumetric flask and made up to the mark with acetonitrile.

For the final preparation of the acetylation reagent, 5 mL 12.5% acetic anhydride solution and 5 mL 12.5% triethylamine solution are pipetted into a 10 mL volumetric flask and are mixed thoroughly.

The acetylation reagent must be freshly prepared on every working day.

2.4 Internal standard

A globin labelled with d_5-epichlorohydrin is used as internal standard (IS). To prepare the IS, 200 µL d_5-epichlorohydrin are added to 20 mL non-smoker's pooled lysate (for preparation see Section 3.1) in a 20 mL screw cap glass tube and shaken on a laboratory shaker at 180 min^{-1} for 4 hours. The globin is then isolated according to Section 3.2.

- Stock solution:
  10 mg of the isolated globin are weighed into a 50 mL volumetric flask, dissolved in formamide (30 min in the ultrasonic bath) and made up to the mark with formamide.
- Working solution:
  1000 µL of the stock solution are pipetted into a 50 mL volumetric flask and made up to the mark with formamide. 100 µL each of this working solution are added to samples, calibration standards and quality controls using pipettes.

2.5 Calibration standards

- Stock solution (1 mM CHPV)
  39.8 mg of the dipeptide standard N-(3-chloro-2-hydroxypropyl)-valine leucine anilide are weighed exactly into a 100 mL volumetric flask and made up to the mark with ethanol.
- Working solution (1 µM CHPV)
  100 µL of the stock solution are pipetted into a 100 mL volumetric flask already containing about 50 mL ethanol and made up to the mark with ethanol.

Starting with this working solution, the spiking solutions for calibration are prepared according to the pipetting scheme in Table 1.

Table 1 Pipetting scheme for the preparation of the spiking solutions for calibration.

Spiking solution	Volume of ethanol [mL]	Volume of stock solution [mL]	Total volume [mL]	CHPV level [nmol]
1	20.0	0.0	20	0
2	19.8	0.2	20	10
3	19.5	0.5	20	25
4	19.0	1.0	20	50
5	18.5	1.5	20	75
6	18.0	2.0	20	100
7	16.5	3.5	20	175
8	15.0	5.0	20	250

To prepare the calibration standards, 100 mg standard pooled globin (from non-smokers) are weighed into 13 mL screw cap glass tubes, dissolved in 3 mL formamide (30 min in the ultrasonic bath) followed by the addition of 100 µL of the corresponding spiking solution and 100 µL of the working solution of the internal standard (Table 2).

Table 2 Pipetting scheme for the preparation of the calibration standards.

Calibration standard	Volume of working solution IS [µL]	Volume of the corresponding spiking solution [µL]	CHPV in the sample [pmol/g globin]
1	100	100	0
2	100	100	10
3	100	100	25
4	100	100	50
5	100	100	75
6	100	100	100
7	100	100	175
8	100	100	250
internal standard	100	0	approx. 100*

* Estimated value, as the IS was prepared in the laboratory by incubation of d_5-epichlorohydrin with pooled globin. The exact adduct formation rate is not known.

3 Specimen collection and sample preparation

3.1 Preparation of the erythrocyte lysate

Blood samples (approx. 5 mL) are collected with an EDTA monovette (7.5 mL nominal volume). Globin isolation of the whole blood should take place within 24 hours. For this purpose, every sample is initially centrifuged at 800 g for 5 min. The filling level of each monovette is marked on the outside with a permanent marker. The supernatant plasma is pipetted off and 0.9% saline solution is added to the erythrocyte sediment up to the mark and the monovette is mixed thoroughly. The sample is centrifuged again at 800 g for 5 min and the supernatant is again pipetted off and discarded. The washing process is repeated until the supernatant is colourless (at least twice). Following this procedure, about 2.5 mL plasma-free erythrocyte sediment is obtained from 5 mL blood. Lysis of erythrocytes is performed by adding bidist. water up to the mark and deep freezing the sample at −20°C for at least eight hours.

3.2 Isolation of the globin

The deep frozen lysate sample is slowly thawed in a water bath at about 37°C and is homogenised by shaking. 2 mL lysate are then transferred with a pipette to a 20 mL screw cap glass tube. After addition of 12 mL 50 mM hydrochloric acid in 2-propanol, the sample is shaken thoroughly and the cell debris are removed by centrifugation (3500 g, 10 min). The supernatant is transferred to a new 20 mL screw cap glass tube and the globin is precipitated by slow addition of 8 mL ethyl acetate.

To complete precipitation, the sample is stored at 4°C in the refrigerator for one hour, followed by centrifugation at 3500 g for 5 min. The supernatant is decanted off and discarded, whereas the precipitated globin is resuspended thrice using 10 mL ethyl acetate each time and is shaken and centrifuged at 3500 g for 5 min each. The supernatant is decanted off and discarded after each centrifugation process. After the final decanting, the precipitated globin is stored overnight in the vacuum desiccator for drying. The dry globin can be stored in glass vials or polypropylene tubes up to 12 months at −27°C without noteworthy adduct losses.

3.3 Derivatisation of the globin

The calibration standard and the internal standard are pipetted according to the scheme in Section 2.5.

From the sample to be analysed, 100 mg globin are weighed into a 13 mL screw cap glass tube and dissolved in 3 mL formamide (30 min in an ultrasonic bath). Then, 100 µL working solution of the IS, 30 µL 1 M sodium hydroxide solution as well as 20 µL pentafluorophenyl isothiocyanate are added to the sample. The calibration standards and quality control standards are processed analogously. After mixing of the sample solution on a laboratory shaker, derivatisation is carried out overnight at room temperature using a horizontal shaker (120 min^{-1}) followed by incubation in a water bath for 2 hours at 45°C.

After equilibration to room temperature, the solution is extracted two times with 4 mL diethyl ether each (samples are mixed thoroughly on a laboratory shaker for 1 min per process). After centrifugation (10 min at 3500 g) the ether phases are transferred to a new 13 mL screw cap glass tube and evaporated to dryness in a stream of nitrogen. The residue is dissolved in 1.5 mL toluene and washed once with 2 mL 0.1 M sodium carbonate solution and twice with 2 mL water. Each washing process includes 1 min mixing on a laboratory shaker and a centrifugation at 3500 g for 5 min. The toluene phase is transferred to a new 13 mL screw cap glass tube after every cleaning step. After the last washing procedure, the toluene phase is transferred as completely as possible into a 1.8 mL glass vial. The sample is evaporated to dryness in a stream of nitrogen at approx. 35°C.

Subsequently, 250 µL acetylation reagent are pipetted into each glass vial. The open sample vials are briefly shaken on a laboratory shaker and then allowed to stand open for 15 min at room temperature. After evaporation to dryness in a stream of nitrogen at 30°C, the residue is dissolved in 500 µL n-hexane and 1000 µL wash solution are added. The sample vial is sealed and mixed thoroughly on a laboratory shaker for about 1 min. After centrifugation (10 min at 3500 g) the hexane phase of the sample is transferred to a new 1.8 mL glass vial and evaporated to dryness in a stream of nitrogen at room temperature. The residue is dissolved in 30 µL toluene, transferred to a 1.8 mL glass vial with a 200 µL micro-insert, sealed and subjected to GC-MS/MS analysis.

4 Operational parameters

4.1 Operational parameters for gas chromatography

Capillary column:	Material:	Quartz glass (fused silica)
	Stationary phase:	Optima-1-MS
	Length:	30 m
	Inner diameter:	0.25 mm
	Film thickness:	0.25 µm
Temperature:	Column:	1 min at 80°C; then increase at 8°C/min to 340°C; 2 min at final temperature
	Injector:	Initial temperature 120°C, then increase at 10°C/s to 320°C, 4 min at final temperature
	Transferline:	300°C
Carrier gas:	Helium 5.0 (constant flow: 1.0 mL/min)	
Sample volume:	1 µL (splitless)	

4.2 Operational parameters for tandem mass spectrometry

Ionisation type:	Negative chemical ionisation (NCI)
Source temperature:	180°C
Reagent gas:	Methane 5.5 (pre-pressure 530 Pa);
Collision gas:	Argon 5.0 (0.27 Pa)
Electron energy	100 eV
Electron flow:	300 µA
Collision energy:	15 eV
Detection:	Selected Reaction Monitoring (SRM)
Measuring time per ion:	100 ms
Electron multiplier:	2800 V

All other parameters must be optimised according to the manufacturer's instructions.

5 Analytical determination

On account of the stereoisomery at the C2 atom of the adduct, the derivatives of the unlabelled 3-chloro-2-hydroxypropyl-valine (d_0-CHPV) and of the deuterated internal standard (d_5-CHPV) eluate as two baseline-separated peaks each with retention times of 21.4 min and 21.6 min, respectively (see chromatogram in Figure 3). The second peak with the longer retention time is the main isomer in each case and was therefore used as reference for quantitative analysis (see Table 3).

Table 3 Retention times of the analytes and ion traces.

Analyte	Retention time [min]	Analysed fragmentation reaction (SRM)
CHPV (analyte)	21.4/**21.6**	$m/z\ 422 \rightarrow 301$
d_5-CHPV (IS)	21.4/**21.6**	$m/z\ 427 \rightarrow 301$

The mass spectra of the CHPV derivatives are shown in Figure 4. The anions with the mass-to-charge ratios (m/z) of m/z 458 and m/z 460 in the mass spectrum of d_0-CHPV (see Figure 4 top row) correspond to the molecule ions with the isotopes ^{35}Cl or ^{37}Cl. The anions with m/z 463 and m/z 465 in the mass spectrum of d_5-CHPV (Figure 4 bottom row) also correspond to the molecule ions with the isotopes ^{35}Cl or ^{37}Cl. Elimination of HCl from these derivatives results in the most intensive ions at m/z 422 (d_0-CHPV) and m/z 427 (d_5-CHPV) (for ^{35}Cl). The loss of H^{35}Cl but not of ^{2}H^{35}Cl indicates that hydrogen atoms from the acetyl function participate in the formation of these ions. The ions at m/z 323 occur in the mass spectra of both derivatives, indicating that this fragment no longer contains the original epichlorohydrin side chain.

To obtain product ion mass spectra, the ions with m/z 458 (for d_0-CHPV) and with m/z 463 (d_5-CHPV) were subjected to a collision-induced dissociation (CID). Figure 5 shows the obtained MS/MS-spectra. The most intensive product ion was obtained at m/z 301 in each case and therefore does not contain an epichlorohydrin side chain.

6 Calibration

The calibration standards are prepared and processed according to Section 2.5 and Section 3 and are analysed as described in Section 4 and Section 5. Calibration graphs are obtained by plotting the quotients of the peak areas of the analyte and the internal standard against the spiked concentration in pmol/g globin. The calibration graph is linear between 10 and 250 pmol/g globin.

7 Calculation of the analytical result

To determine the analyte concentration in a sample, the peak area of the analyte is divided by the peak area of the internal standard. The quotient thus obtained is inserted into the corresponding calibration graph (see Section 6). The CHPV concentration level of the sample is obtained in pmol/g globin. Use the following algorithm to convert into the unit µg/L blood (c = concentration):

$$c \,[\mu g/L \text{ blood}] = c \,[pmol/g \text{ globin}] \times (209 \times 10^{-6} \,\mu g/pmol) \times 144 \,g/L.$$

8 Standardisation and quality control

Quality control of the analytical results is carried out as stipulated in the guidelines of the *Bundesärztekammer* (German Medical Association) [16]. As the adducts are not stable when dissolved, either pooled globin must be prepared from an excess of unspiked globin and a small quantity of labelled globin and kept dry (for use as quality control material), or the slope of the calibration function determined on a daily basis is used as a parameter in quality control.

9 Evaluation of the method

9.1 Precision

To determine the within day precision ten calibration standards with a CHPV concentration of 25 pmol/g globin or 100 pmol/g globin were prepared and analysed in a row. The relative standard deviations were 12.4% (prognostic range: 27.6%, mean value: 31.8 ± 3.9 pmol/g globin) or 9.8% (prognostic range: 21.8%, mean value: 93.7 ± 9.2 pmol/g globin), respectively. The precision from day to day was determined on the basis of the variances of the slopes of ten four-point calibration curves (0, 25, 100, 250 pmol/g globin) and was 15.0% (prognostic range: 33.4%).

9.2 Accuracy

The accuracy of the method was determined to be 99.8 ± 12.5% (range: 84–129%) and was calculated as relative recovery rate of calibration standards with a CHPV concentration of 100 pmol/g globin (n = 10).

9.3 Detection limit and quantitation limit

According to the DIN method 32645 (German Industrial Standard [*Deutsche Industrie-Norm*]) [17], the detection limit of the procedure (calibration graph method) was determined as 10 pmol/g globin, corresponding to about 0.3 µg CHPV per litre blood. The quantitation limit obtained accordingly is 25 pmol/g globin (0.75 µg CHPV per litre blood).

9.4 Sources of error

Care must be taken that the toluene phase is completely evaporated to dryness prior to the acetylation step, as traces of water considerably impair derivatisation.

In the course of storage experiments, it was observed that the protein adduct is not stable in solution. This concerns both, the decrease in adduct concentration in the material to be analysed (haemolysate) and in the calibration standards. As regards the storage of the material to be analysed, it is therefore necessary, to carry out isolation of the globin and its subsequent storage in the freezer as soon as possible after specimen collection. Currently, there are no indications that a degradation of the adduct occurs under these conditions. Preparation and processing of the calibration standards on a day-to-day basis is recommended.

10 Discussion of the method

The so-called modified Edman degradation according to Törnqvist et al. [18] is a suitable reaction for the single step release of terminal epichlorohydrin-valine adducts from globin and the chemical conversion of N-(3-chloro-2-hydroxypropyl)-valine to a derivative with properties well suited for analysis using GC-MS/MS. To increase the thermic stability and to improve the volatility of the Edman derivative its hydroxyl group is acetylated in a second derivatisation step. On account of the pentafluorophenyl function, the derivative obtained is a very strong electron withdrawing molecule and therefore well suited to form anions by negative chemical ionisation (NCI). For perfluorated analytes this ionisation method is clearly more sensitive compared to electron impact ionisation [19, 20]. The quantitative analysis of N-(3-chloro-2-hydroxypropyl)-valine as pentafluorophenyl-thiohydantoin derivative under MS/MS conditions is characterised by very high specifity and sensitivity. By detection of characteristic product ions, formed from precursor ions, which contain the characteristic chlorine atom, interference by co-elution of other substances is practically excluded. Therefore, a qualifier ion is not necessarily required. The use of GC-MS/MS instead of GC-MS provides an additional gain in sensitivity due to a lower background noise so that a low detection limit of 10 pmol/g globin is obtained. The quantitative determination of N-(3-chloro-2-hydroxypropyl)-valine as pentafluorophenyl thiohydantoin derivative by means of GC-MS/MS using pentadeuterated epichlorohydrin as internal standard is sufficiently precise in the low pmol/g range. On account of these characteristics, the present procedure is well suited for the detection and quantitation of globin adducts of epichlorohydrin in blood.

11 References

1 BUA (Beratergremium für umweltrelevante Altstoffe): Epichlorhydrin. BUA-Stoffbericht 90, Wiley-VCH Verlag, Weinheim (1992).
2 H. Greim (ed.): 1-Chlor-2,3-epoxypropan (Epichlorhydrin). Gesundheitsschädliche Arbeitsstoffe, Toxikologisch-arbeitsmedizinsche Begründungen für MAK-Werte. Volume 36, Wiley-VCH Verlag, Weinheim (2003).
3 International Agency for Research on Cancer (IARC): Re-evaluation of Some Organic Chemicals, Hydrazine and Hydrogen Peroxide: Epichlorhydrin. Monographs on the Evaluation of Carcinogenic Risks to Humans. Volume 71, 603–628 (1999).
4 W.W. Weigel, H.B. Plotnick, W.L. Conner: Tissue distribution and excretion of 14C-epichlorohydrin in male and female rats. Res. Commun. Chem. Pathol. Pharmacol. 20, 275–287 (1978).
5 R. Gingell, H.R. Mitschke, I. Dzidic, P.W. Beatty, V.L. Sawin, A.C. Page: Disposition and metabolism of [2-14C]epichlorohydrin after oral administration to rats. Drug Metab. Dispos. 13, 333–341 (1985).
6 B.M. De Rooij, J.N.M. Commandeur, J.R. Ramcharan, H.C. Schuilenburg, B.L. van Baar, N.P.E. Vermeulen: Identification and quantitative determination of 3-chloro-2-hydroxypropylmercapturic acid and alpha-chlorohydrin in urine of rats treated with epichlorohydrin. J. Chromatogr. B Biomed. Appl. 685, 241–250 (1996).
7 B.M. De Rooij, P.J. Boogaard, J.N.M. Commandeur, N.P.E. Vermeulen: 3-Chloro-2-hydroxypropylmercapturic acid and alpha-chlorohydrin as biomarkers of occupational exposure to epichlorohydrin. Environ. Tox. Pharmacol. 3, 175–185 (1997).
8 H. Hindsø Landin, T. Grummt, C. Laurent, A. Tates: Monitoring of occupational exposure to epichlorohydrin by genetic effects and hemoglobin adducts. Mutat. Res. 381, 217–226 (1997).
9 H. Hindsø Landin, D. Segerbäck, C. Damberg, S. Osterman-Golkar: Adducts with haemoglobin and with DNA in epichlorohydrin-exposed rats. Chem. Biol. Interact. 117, 49–64 (1999).
10 N. Miraglia, G. Pocsfalvi, P. Ferranti, A. Basile, N. Sannolo, A. Acampora, L. Soleo, F. Palmieri, S. Caira, B. De Giulio, A. Malorni: Mass spectrometric identification of a candidate biomarker peptide from the in vitro interaction of epichlorohydrin with red blood cells. J. Mass Spectrom. 36, 47–57 (2001).
11 J. Mäki, K. Karlsson, R. Sjöholm, L. Kronberg: Structural characterisation of the main epichlorohydrin-guanosine adducts. Adv. Exp. Med. Biol. 500, 125–128 (2001).
12 K. Plna, S. Osterman-Golkar, E. Nogradi, D. Segerbäck: 32P-postlabelling of 7-(3-chloro-2-hydroxypropyl)guanine in white blood cells of workers occupationally exposed to epichlorohydrin. Carcinogenesis 21, 275–280 (2000).
13 M. Bader, W. Rosenberger, R. Wrbitzky, F.M. Gutzki, D. Tsikas, D.O. Stichtenoth: Biomonitoring von Einsatzkräften nach akzidenteller Exposition gegenüber Epichlorhydrin – Ergebnisse und Risikoabschätzung. Tagungsband zur 48. Jahrestagung der Deutschen Gesellschaft für Arbeitsmedizin und Umweltmedizin (DGAUM), S. 471–474 Hamburg (2008).
14 H. Hindsø Landin, S. Osterman-Golkar, V. Zorcec, M. Törnqvist: Biomonitoring of epichlorohydrin by hemoglobin adducts. Anal. Biochem. 240, 1–6 (1996).

15 K.M. Wollin, M. Bader, M. Müller, W. Lilienblum, M. Csicsaky: Assessment of long-term health effects by means of haemoglobin adducts of 1-chloro-2,3-epoxypropane (ECH) after accidental exposure. Naunyn-Schmiedeberg's Arch. Pharmacol. 377, Supplement 1, 92 (2008).

16 Bundesärztekammer: Richtlinie der Bundesärztekammer zur Qualitätssicherung quantitativer laboratoriumsmedizinischer Untersuchungen. Dt. Ärztebl. 100, A3335–A3338 (2003).

17 Deutsche Industrienorm (DIN): No. 32645: Nachweis-, Erfassungs- und Bestimmungsgrenze. Beuth Verlag GmbH, Berlin (1994).

18 M. Törnqvist, J. Mowrer, S. Jensen, L. Ehrenberg: Monitoring of envirnonmental cancer initiators through hemoglobin adducts by a modified Edman degradation method. Anal. Biochem. 154, 255–266 (1986).

19 D. Tsikas, K. Caidahl: Recent methodological advances in the mass spectrometric analysis of free and protein-associated 3-nitrotyrosine in human plasma. J. Chromatogr. B. 814, 1–9 (2005).

20 D. Tsikas: Application of gas chromatography-mass spectrometry and gas chromatography-tandem mass spectrometry to assess in vivo synthesis of prostaglandins, thromboxane, leukotrienes, isoprostanes and related compounds in humans. J. Chromatogr. B. 717, 201–245 (1998).

Authors: *M. Bader, W. Rosenberger, D. Tsikas, F.M. Gutzki*
Examiners: *Th. Göen, E. Eckert*

12 Appendix

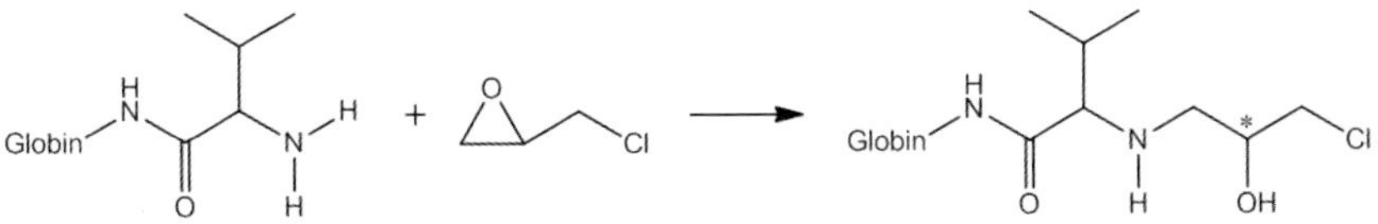

Fig. 1 Formation of the primary adduct N-(3-chloro-2-hydroxypropyl)-valine (CHPV) from the reaction of epichlorohydrin with the N-terminal amino acid of a globin chain.

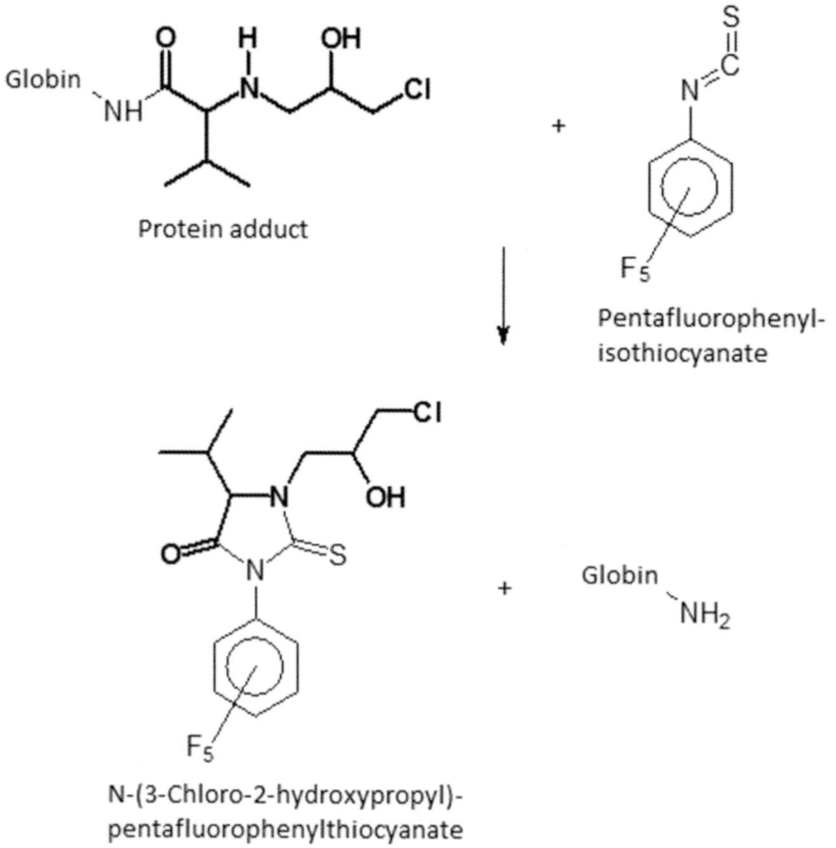

Fig. 2 Cleavage and derivatisation of N-(3-chloro-2-hydroxypropyl)-valine by modified Edman degradation.

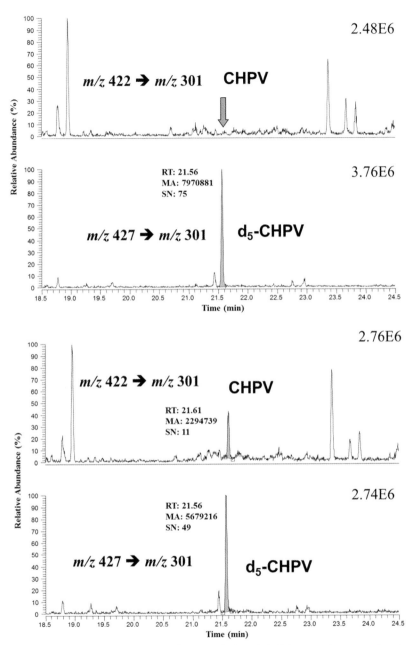

Fig. 3 GC-MS/MS chromatograms of an unspiked pooled globin sample (top row) and of a pooled globin sample spiked with 25 pmol CHPV/g globin (bottom row). The concentration level of d_5-CHPV in both samples was approx. 100 pmol/g globin.

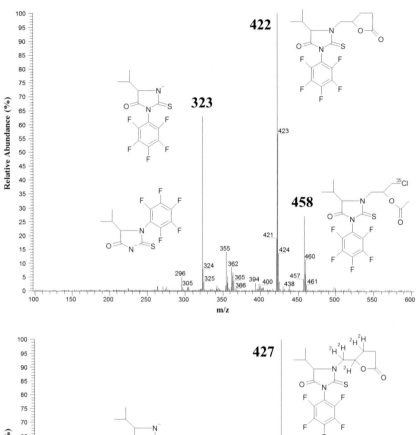
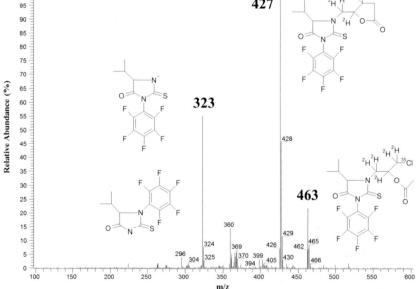

Fig. 4 Mass spectra of d_0-CHPV (top) and d_5-CHPV (bottom).

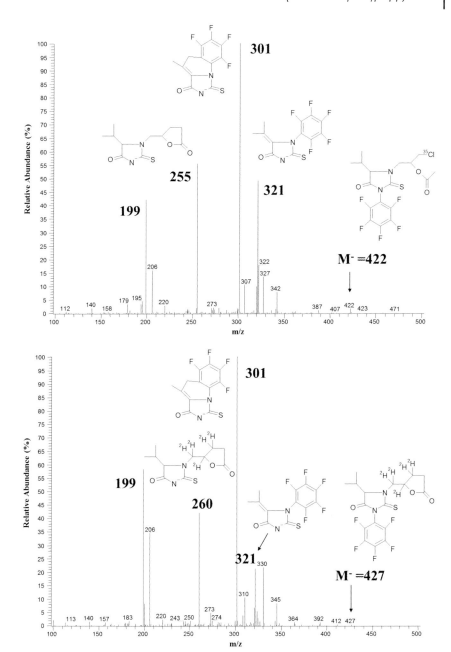

Fig. 5 MS/MS spectra of d_0-CHPV (top) and d_5-CHPV (bottom).

Cyclohexanol, 1,2-Cyclohexanediol and 1,4-Cyclohexanediol in Urine

Matrix:	Urine
Hazardous substances:	Cyclohexane, cyclohexanone and cyclohexanol
Analytical principle:	Capillary gas chromatography/flame ionisation detection (GC/FID)
Completed in:	June 2001

Overview of parameters that can be determined with this method and the corresponding hazardous substances:

Hazardous substance	CAS	Parameter	CAS
Cyclohexane	110-82-7	Cyclohexanol	108-93-0
		1,2-Cyclohexanediol	931-17-9
		1,4-Cyclohexanediol	556-48-9
Cyclohexanone	108-94-1	Cyclohexanol	108-93-0
		1,2-Cyclohexanediol	931-17-9
		1,4-Cyclohexanediol	556-48-9
Cyclohexanol	108-93-0	Cyclohexanol	108-93-0
		1,2-Cyclohexanediol	931-17-9
		1,4-Cyclohexanediol	556-48-9

Summary

The method described below permits the determination of cyclohexanol, 1,2-cyclohexanediol and 1,4-cyclohexanediol in urine as biomarkers of exposure to cyclohexane, cyclohexanone and cyclohexanol.

After addition of the internal standard n-tridecane, the urine is subjected to hydrolysis using hydrochloric acid. The urine is neutralised with potassium carbonate. Subsequently, the analytes are enriched by means of liquid-liquid extraction, while being simultaneously separated from the matrix components. After separation by capillary gas chromatography, the extract is analysed using a flame ionisation detector.

Calibration curves are plotted for quantitative analysis using calibration standard solutions prepared by spiking pooled human urine with the standard substances. For analysis, the calibration standard solutions are treated in the same manner as the urine samples.

Cyclohexanol

Within day precision:	Standard deviation (rel.)	s_w = 8.7% or 7.5%
	Prognostic range	u = 19.4% or 16.7%
	at a spiked concentration of 5 mg or 20 mg cyclohexanol per litre urine and where n = 10 determinations	
Day to day precision:	Standard deviation (rel.)	s_w = 9.4% or 8.0%
	Prognostic range	u = 20.5% or 17.4%
	at a spiked concentration of 5 mg or 20 mg cyclohexanol per litre urine and where n = 12 determinations	
Accuracy:	Recovery rate	r = 98.1% or 101.2%
	at nominal concentrations of 5 mg or 20 mg cyclohexanol per litre urine and where n = 10 determinations	
Detection limit:	1 mg cyclohexanol per litre urine	
Quantitation limit:	3 mg cyclohexanol per litre urine	

1,2-Cyclohexanediol

Within day precision:	Standard deviation (rel.)	s_w = 6.4% or 7.0%
	Prognostic range	u = 14.3% or 15.6%
	at a spiked concentration of 20 mg or 80 mg 1,2-cyclohexanediol per litre urine and where n = 10 determinations	
Day to day precision:	Standard deviation (rel.)	s_w = 6.8% or 7.4%
	Prognostic range	u = 14.8% or 16.1%
	at a spiked concentration of 20 mg or 80 mg 1,2-cyclohexanediol per litre urine and where n = 12 determinations	
Accuracy:	Recovery rate	r = 96.0% or 99.8%
	at nominal concentrations of 25 mg or 100 mg 1,2-cyclohexanediol per litre urine and where n = 10 determinations	
Detection limit:	1 mg 1,2-cyclohexanediol per litre urine	
Quantitation limit:	3 mg 1,2-cyclohexanediol per litre urine	

1,4-Cyclohexanediol

Within day precision:	Standard deviation (rel.)	s_w = 7.3% or 7.9%
	Prognostic range	u = 16.3% or 17.6%
	at a spiked concentration of 20 mg or 80 mg 1,4-cyclohexanediol per litre urine and where n = 10 determinations	

| Day to day precision: | Standard deviation (rel.) | S_w = 8.0% or 7.5% |
| | Prognostic range | u = 17.4% or 16.3% |

at a spiked concentration of 20 mg or 80 mg 1,4-cyclohexanediol per litre urine and where n = 12 determinations

Accuracy: Recovery rate r = 84.1% or 97.6%

at nominal concentrations of 25 mg or 100 mg 1,4-cyclohexanediol per litre urine and where n = 10 determinations

Detection limit: 1 mg 1,4-cyclohexanediol per litre urine
Quantitation limit: 3 mg 1,4-cyclohexanediol per litre urine

Cyclohexanol, 1,2-cyclohexanediol and 1,4-cyclohexanediol in urine

Cyclohexane is a colourless, highly inflammable liquid. It is insoluble in water, but readily soluble in alcohols, hydrocarbons, ethers or chlorinated hydrocarbons. Cyclohexane is used mainly as an intermediate in the production of nylon. It is used in the production of cyclohexanol and cyclohexanone, rubber and glues, adipic acid, ε-caprolactam and hexamethylenediamine. In addition, it is used as a solvent for lacquers and resins, for the extraction of essential oils and also as a solvent in spectroscopy [1–3]. Cyclohexane exposure occurs mainly by inhalation; a large proportion is exhaled unchanged [4]. Pulmonary retention of the substance was given as 18.4% (resting conditions) and 34% (physical exercise) [5].

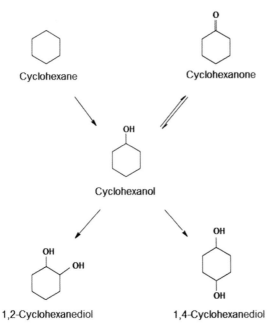

Fig. 1 Metabolism of cyclohexane in humans [6].

The metabolism of cyclohexane as known from field studies and experimental studies with humans is presented in Figure 1. Accordingly, the presented method is also suitable for determination of occupational exposure to cyclohexanone and cyclohexanol.

The metabolites cyclohexanol, 1,2-cyclohexanediol and 1,4-cyclohexanediol are excreted by humans both as glucuronides and as free non-conjugates [6]. All three cyclohexane metabolites are suitable as possible biomarkers of exposure to cyclohexane; although, the BAT value (*Biologischer Arbeitsstoff-Toleranzwert*, biological tolerance value) refers to 1,2-cyclohexanediol only. In the addendum to cyclohexane, the new BAT value was established at 150 mg 1,2-cyclohexanediol per g creatinine after hydrolysis (sampling time: end of exposure or end of shift and, for long-term exposure, after several previous shifts) [7, 8]. Moreover, a EKA correlation (*Expositionsäquivalente für krebserzeugende Arbeitsstoffe*, exposure equivalents for carcinogenic substances) was published for cyclohexanol in urine and 1,2-cyclohexanediol in urine as biomonitoring parameters of cyclohexane.

After exposure to a cyclohexane concentration level of 1050 mg/m^3, conducted with volunteers over a period of 72 hours, urinary analysis showed that 0.5% was excreted in the urine as cyclohexanol, 23.4% as 1,2-cyclohexanediol and 11.3% as 1,4-cyclohexanediol (determined as the sum of the free and conjugated compounds). Metabolism to cyclohexanone could not be demonstrated in this study [9]. Intermediate dehydroxylation to benzene, which was assumed by some authors, does not take place [10, 11].

As a result of further oxidative biotransformation, urinary cyclohexanol levels are considerably lower than urinary cyclohexanediol levels. The excretion half-life of 1,2-cyclohexanediol was found to be on average 17.0 ± 5.2 hours [12] and 15.8 ± 1.9 hours [6, 13]. Mráz et al. [12] determined a similar half-life for 1,4-cyclohexanediol, while Trabs [6] reported a half-life that was around twice as high, with an average value of 31.5 hours. The elimination of cyclohexanol, however, was found to be comparatively rapid, with a mean half-life of 4.6 hours [6, 13] and 1.5 hours [12]. For the monitoring of occupationally exposed persons, the cyclohexane metabolites 1,2-cyclohexanediol and 1,4-cyclohexanediol should therefore be determined at the end of the shift after several previous shifts [8].

The toxicity of cyclohexane has been reported in detail in the 1996 MAK value documentation [14]. Cyclohexane is not normally found in the environment or in household products. However, it is a component of some glues and resins. The 1985/86 indoor survey of the German Federal Environmental Agency based on 479 indoor determinations gave a median cyclohexane level of 6 µg/m^3 and a 95th percentile of 18 µg/m^3. In urine samples of the general population, cyclohexane metabolites were not detected with common analytical methods [15]. Using a sensitive method, Perico et al. [16] determined a mean level of 0.5 mg 1,2-cyclohexanediol per g creatinine and a median level of 0.3 mg 1,2 cyclohexanediol per g creatinine in the urine of 31 non-exposed workers.

Author: *U. Knecht*
Examiners: *R. Heinrich-Ramm, K. H. Tieu*

Cyclohexanol, 1,2-Cyclohexanediol and 1,4-Cyclohexanediol in Urine

Matrix:	Urine
Hazardous substances:	Cyclohexane, cyclohexanone and cyclohexanol
Analytical principle:	Capillary gas chromatography/flame ionisation detection (GC/FID)
Completed in:	June 2001

Content

1	General principles
2	Equipment, chemicals and solutions
2.1	Equipment
2.2	Chemicals
2.3	Solutions
2.4	Internal standard
2.5	Calibration standards
3	Specimen collection and sample preparation
4	Operational parameters
4.1	Operational parameters for gas chromatography
5	Analytical determination
6	Calibration
7	Calculation of the analytical result
8	Standardisation and quality control
9	Evaluation of the method
9.1	Precision
9.2	Accuracy
9.3	Detection limits and quantitation limits
9.4	Sources of error
10	Discussion of the method
11	References
12	Appendix

1 General principles

Following an addition of the internal standard n-tridecane, urinary cyclohexanol, 1,2-cyclohexanediol and 1,4-cyclohexanediol are subjected to hydrolysis by hydrochloric acid. The urine is then neutralised and saturated with potassium carbonate. The analytes are subsequently enriched by means of a liquid-liquid extraction using ethylacetate. Without any further enrichment and after separation by capillary gas chromatography, the metabolites are determined using a flame ionisation detector. Calibration is carried out with spiked urine samples processed and analysed in the same way as native urine samples.

2 Equipment, chemicals and solutions

2.1 Equipment

- Gas chromatograph with flame ionisation detector (FID), data processing system, autosampler or 5 µL syringe for the gas chromatograph
- Capillary gas chromatographic column: length: 50 m; inner diameter: 0.53 mm; film thickness: 2 µm, stationary phase: 14%-cyanopropylphenyl-86%-dimethyl-polysiloxane (e.g. CP-Sil 19 CB from Agilent Technologies)
- 250 mL Urine containers with screw cap (e.g. Sarstedt No. 77.577)
- Analytical balance (e.g. Sartorius)
- 20 mL Crimp cap vials with PTFE (polytetrafluoroethylene) septa and aluminium crimp caps as well as a manual crimper (e.g. Macherey-Nagel)
- Thermostatic water bath
- 10 mL and 30 mL Centrifuge tubes (e.g. Sarstedt)
- 10 mL, 20 mL and 100 mL Volumetric flasks
- Piston-stroke pipettes with adjustable volume between 10 µL and 100 µL or 100 µL and 1000 µL with suitable pipette tips (e.g. Eppendorf)
- Finnpipette 1000–5000 µL (e.g. Thermo Scientific)
- Laboratory centrifuge
- 2 mL Screw cap vials and screw caps with PTFE septa
- Laboratory shaker

2.2 Chemicals

- Acetone p.a. (e.g. Merck, No. 100012)
- Cyclohexanol p.a. (e.g. Aldrich, No. 105899)
- cis-1,2-Cyclohexanediol puriss. (e.g. Aldrich, No. 28995)

- trans-1,2-Cyclohexanediol 98% (e.g. Aldrich, No. 141712)
- 1,4-Cyclohexanediol 99% (e.g. Aldrich, No. C101206)
- Ethylacetate p.a. (e.g. Merck, No. 110972)
- Potassium carbonate p.a. (e.g. Merck, No. 104928)
- Sodium sulfate p.a. (e.g. Merck, No. 106637)
- Hydrochloric acid 37% p.a. (e.g. Merck, No. 100317)
- n-Tridecane p.a. (e.g. Merck, No. 109609)
- Bidistilled water
- Helium 5.0 (e.g. Linde)

2.3 Solutions

- Saturated potassium carbonate solution
  100 mL bidistilled water are placed in a beaker. Subsequently, under stirring, potassium carbonate is added (about 112 g) until a sediment of undissolved potassium carbonate remains at the bottom of the beaker.

2.4 Internal standard

- Stock solution (10 g n-tridecane/L)
  100 mg n-tridecane are weighed into a 10 mL volumetric flask, dissolved in acetone and made up to the mark with acetone.
- Working solution (2,5 g n-tridecane/L)
  5 mL of the stock solution are pipetted into a 20 mL volumetric flask. The flask is made up to the mark with acetone.

2.5 Calibration standards

- Stock solution cyclohexanol (1 g cyclohexanol/L)
  100 mg cyclohexanol are weighed into a 100 mL volumetric flask. The flask is made up to the mark with acetone.
- Working solution 1,2-cyclohexanediol and 1,4-cyclohexanediol (10 g 1,2-cyclohexanediol and 10 g 1,4-cyclohexanediol/L)
  50 mg cis-1,2-cyclohexanediol, 50 mg trans-1,2-cyclohexanediol and 100 mg cis-/trans-1,4-cyclohexanediol are weighed into a 10 mL volumetric flask and dissolved in acetone. The flask is made up to the mark with acetone.

Calibration standard solutions are prepared by pipetting between 50 µL and 1 mL of the working solution into a 50 mL volumetric flask and filling it up to the mark with pooled urine. This gives solutions with 1 mg to 20 mg cyclohexanol per litre urine or 10 mg to 200 mg 1,2-cyclohexanediol and 1,4-cyclohexanediol per litre urine, respectively (Table 1).

When stored at –18°C, the calibration solutions are stable for at least 6 months.

Table 1 Pipetting scheme for the preparation of cyclohexanol, 1,2-cyclohexanediol and 1,4-cyclohexanediol calibration standards in urine.

Volume of the working solution [mL]	Final volume of the calibration standard solution [mL]	Analyte level [mg/L]
Cyclohexanol		
1	50	20.0
0.75	50	15.0
0.5	50	10.0
0.25	50	5.0
0.125	50	2.5
0.05	50	1.0
1,2-Cyclohexanediol and 1,4-cyclohexanediol		
1	50	200
0.75	50	150
0.5	50	100
0.25	50	50
0.125	50	25
0.05	50	10

3 Specimen collection and sample preparation

The urine is collected in sealable plastic bottles and should be stored at –18°C until analysis is performed. Under these storage conditions, the urine is stable for at least one year.

The urine sample is thawed in a water bath at 60°C. After equilibration to room temperature and thoroughly mixing, 5 mL urine is pipetted into a 20 mL crimp cap vial and 200 µL of the working solution of the internal standard (2.5 g/L) is added. Following acidification of the sample with 1 mL 37% hydrochloric acid, the crimp cap vial is sealed and the solution subjected to hydrolysis in the water bath at 100°C for one hour. Immediately after cooling to room temperature, the hydrolysate is neutralised by slowly adding a saturated potassium carbonate solution (about 1.3 mL); subsequently, a further 5 g potassium carbonate are added in solid form. After mixing with 2 mL ethylacetate the crimp cap vial is sealed again and thoroughly shaken for 20 min. For centrifugation (4000 U/min), the sample is transferred to a centrifuge tube. Subsequently, a 1 mL aliquot of the ethylacetate phase is transferred into a screw cap vial and dried over sodium sulfate.

4 Operational parameters

4.1 Operational parameters for gas chromatography

Capillary column:	Material:	Fused Silica
	Stationary phase:	CP-SiI 19 CB
	Length:	50 m
	Inner diameter:	0.53 mm
	Film thickness:	2.0 µm
Detector:	Flame ionisation detector (FID)	
Temperature:	Column:	Initial temperature 160°C, 10 min isothermal, then increase at a rate of 30°C/min to 220°C; 8 min isothermal; then increase to 250°C at 20°C/min
	Injector:	220°C
	Detector:	260°C
Carrier gas:	Helium 5.0 at a column pre-pressure of 50 kPa	
Split:	1:10	
Injection volume:	1 µL	

All other parameters must be optimised in accordance with the manufacturer's instructions.

5 Analytical determination

The urine samples processed as described in Section 3, are analysed by injecting 1 µL of the ethylacetate extract into the GC/FID system. At least one urine control sample and one reagent blank are analysed with each series. In the case of the latter, bidistilled water is used in place of urine.

Figure 2 shows the chromatogram of a urine sample of a volunteer exposed to cyclohexane. The retention times shown in Figure 2 for cyclohexanol, 1,2-cyclohexanediol and 1,4-cyclohexanediol are intended to be a rough guide only. Users of the method must ensure proper separation power of the capillary column used and the resulting retention behaviour of the analytes.

6 Calibration

The calibration standard solutions in urine prepared according to Section 2.5 are processed in the same way as the urine samples (see Section 3), and analysed by gas chromatography with a flame ionisation detector (see Section 4). Calibration curves are obtained by plotting the quotients of the peak areas of cyclohexanol, 1,2-cyclohexanediol and 1,4-cyclohexanediol to the peak area of the internal standard as a function of the spiked concentration. It is not necessary to plot a complete calibration graph for every analytical series. As a rule, it is sufficient to include one calibration standard in every analytical series. In that case, the peak area ratio of the one calibration standard is calculated and compared to the peak area ratio of the equivalent standard in the complete calibration function. A new calibration graph should be plotted if the quality control results indicates a systematic deviation. The calibration graph for cyclohexanol is linear between 1 mg and 20 mg per litre urine. For 1,2-cyclohexanediol and 1,4-cyclohexanediol the calibration graph is linear between 10 mg and 200 mg each per litre urine. Figures 3 to 5 show examples of the linear calibration graphs of the three metabolites in urine.

7 Calculation of the analytical result

The analyte concentration level in the urine samples is calculated on the basis of the calibration graph obtained according to Section 6. The peak area of each analyte is divided by the peak area of the internal standard. From the quotient thus obtained, the corresponding concentration of the three analytes in mg per litre urine is obtained using the relevant calibration graph. A possible reagent blank value has to be taken into consideration.

8 Standardisation and quality control

Quality control of the analytical results is carried out as stipulated in the guidelines of the *Bundesärztekammer* (German Medical Association) [17]. For quality control, urine spiked with a defined amount of the analytes is processed within each analytical series. A quality control material for cyclohexanol, 1,2-cyclohexanediol and 1,4-cyclohexanediol in urine is not commercially available. Thus, it has to be prepared in the laboratory. For this purpose, pooled urine of individuals occupationally non-exposed to cyclohexane, cyclohexanone or cyclohexanol is spiked with a defined quantity of cyclohexanol, 1,2-cyclohexanediol and 1,4-cyclohexanediol. The obtained concentration level should lie within the relevant concentration range. The quality control urine is divided into aliquots. Stored at −18°C, it can be used for up to 12 months. The nominal value and the tolerance range of this quality control material are determined in a pre-analytical period (one analysis of the control material on 15 different days) [17–19].

9 Evaluation of the method

9.1 Precision

The precision was determined using pooled urine samples of persons occupationally non-exposed to cyclohexane. These samples were spiked with defined levels of cyclohexanol as well as 1,2-cyclohexanediol and 1,4-cyclohexanediol leading to a final concentration of 5 mg/L and 20 mg/L for cyclohexanol and 20 mg/L and 80 mg/L for each of the two cyclohexanediols. Within day precision was determined by analysis of these samples ten times in a row and yielded the relative standard deviations and the corresponding prognostic ranges given in Table 2.

Table 2 Within day precision for the determination of cyclohexanol, 1,2-cyclohexanediol and 1,4-cyclohexanediol in urine (n = 10).

Analyte	Spiked concentration [mg/L]	Standard deviation (rel.) s_w [%]	Prognostic range u [%]
Cyclohexanol	5	8.7	19.4
	20	7.5	16.7
1,2-Cyclohexanediol	20	6.4	14.3
	80	7.0	15.6
1,4-Cyclohexanediol	20	7.3	16.3
	80	7.9	17.6

Furthermore, day to day precision was determined by analysis of the spiked urine samples on twelve different days. The resulting standard deviations are shown in Table 3.

Table 3 Day to day precision for the determination of cyclohexanol, 1,2-cyclohexanediol and 1,4-cyclohexanediol in urine (n = 12).

Analyte	Spiked concentration [mg/L]	Standard deviation (rel.) s_w [%]	Prognostic range u [%]
Cyclohexanol	5	9.4	20.5
	20	8.0	17.4
1,2-Cyclohexanediol	20	6.8	14.8
	80	7.4	16.1
1,4-Cyclohexanediol	20	8.0	17.4
	80	7.5	16.3

9.2 Accuracy

The accuracy of the method and the analyte losses occurring during sample preparation were assessed using pooled urine spiked with the analytes at two different concentration levels and dividing it into aliquots. Each urine sample was processed and analysed 10 times as described in Section 3. The mean relative recovery rates were found to range between 84% and 101% (Table 4).

Table 4 Recovery rates for cyclohexanol, 1,2-cyclohexanediol and 1,4-cyclohexanediol in urine (n = 10).

Analyte	Concentration after spiking [mg/L]	Recovery rate r [%]	Range [%]
Cyclohexanol	5	98.1	94.7–101.2
	20	101.2	97.9–104.0
1,2-Cyclohexanediol	25	96.0	91.7–100.4
	100	99.8	94.6–107.7
1,4-Cyclohexanediol	25	84.1	81.9–96.2
	100	97.6	92.0–109.2

Analyte losses due to sample preparation were determined by analysing the spiked urine samples five times in a row. Subsequently, the obtained mean peak areas were compared to the peak areas of the analytes obtained by the analysis of respective standard solutions in acetone. The losses due to sample preparation lie within the range of 9% to 22%.

9.3 Detection limits and quantitation limits

The limits of detection of the three analytes have been calculated to be 1 mg per litre urine (based on a signal to noise ratio of 1:3). This corresponds to quantification limits of 3 mg per litre urine of each analyte.

9.4 Sources of error

As described in Section 3, the neutralisation of the hydrolysate with potassium carbonate should progress slowly as strong gas evolution resulting from a fast addition of potassium carbonate can cause analyte losses.

The occurrence of interfering peaks was occasionally observed for 1,2-cyclohexanediol, leading to false-positive results. For example, the examiner of the method observed a signal at the retention time of 1,2-cyclohexanediol in a number of urine samples from occupationally non-exposed persons. With the detector used (FID) he was not able to clarify whether the signal was really caused by 1,2-cyclohexanediol.

10   Discussion of the method

The described analytical method is a suitable and validated procedure for the determination of cyclohexanol, 1,2-cyclohexanediol and 1,4-cyclohexanediol in urine. The method allows the simultaneous determination of the metabolites measuring the total (free and conjugated) analyte level in urine.

An essential advantage of determining cyclohexane metabolites in urine is the fact that the metabolite concentrations in urine are proportional to external exposure. In the case of occupational exposure, the urinary concentration levels of the cyclohexane metabolites, in particular those of the diols, are comparatively high, while cyclohexanol and the isomeric diols are not detectable in the urine of persons occupationally non-exposed to cyclohexane. Cyclohexanol, 1,2-cyclohexanediol and 1,4-cyclohexanediol can also be used as biomarkers of occupational exposure to cyclohexanone and cyclohexanol.

As the metabolically formed cis/trans-isomers of 1,2-cyclohexanediol and 1,4-cyclohexanediol cannot be separated or only separated in negligible amounts, this method measures the total level of both isomers.

Calibration should be carried out in pooled urine, since matrix effects cannot be definitely ruled out in the case of calibration curves prepared in water. Background exposure levels above the detection limit are usually not observed when urine of occupationally non-exposed persons is used for the preparation of pooled urine. The method given permits a reliable determination of cyclohexanol, 1,2-cyclohexanediol and 1,4-cyclohexanediol in the urine after exposure at the workplace. The method is easy to handle, robust and uses a common analytical technique (GC-FID). The reliability criteria of this method can be considered as good. The same is true for the linearity of the calibration graph using the saturated hydrocarbon n-tridecane as an internal standard. However, the linearity may be further improved using an alcohol as internal standard instead.

Instruments used:

- Hewlett Packard Capillary Gas Chromatograph A 5890 with flame ionisation detector and autosampler.

11   References

[1] F. K. Beilstein (ed.): Beilsteins Handbuch der Organischen Chemie. Springer Berlin, 4th Edition; E IV, 27–44 (1985).

2 R. E. Kirk and D. F. Othmer (eds.): Encyclopedia of Chemical Technology. Wiley-Interscience New York, 4th Edition; 13, 829–835 (1998).
3 K. Weissermel and H. Arpe (eds.): Industrielle organische Chemie. VCH Weinheim, 4th Edition; 374 (1994).
4 E. E. Sandmeyer: Alicyclic hydrocarbons. In: G. D. Clayton and E. E. Clayton (eds.) Patty's Industrial Hygiene and Toxicology. Wiley-Interscience New York, 3th Edition; 3221–3225 (1981).
5 A. Mutti, M. Falzoi, S. Lucertini, A. Cavatorta, I. Franchini, C. Pedroni: Absorption and alveolar excretion of cyclohexane in workers in a shoe factory. J. Appl. Toxicol. 1, 220–223 (1981).
6 A. Trabs: Biologisches Monitoring nach standardisierter Cyclohexan-Einwirkung. Inaugural-Dissertation des Fachbereiches Humanmedizin der JLU Giessen (1999).
7 H. Greim and H. Drexler (eds.): Cyclohexane. The MAK-Collection Part II: BAT Value Documentations. Volume 4, Wiley-VCH Verlag, Weinheim (2005).
8 A. Hartwig and H. Drexler (eds.): Addendum zu Cyclohexan. Biologische Arbeitsstoff-Toleranz-Werte (BAT-Werte) und Expositionsäquivalente für krebserzeugende Arbeitsstoffe (EKA) und Biologische Leitwerte (BLW). 17th issue, Wiley-VCH, Weinheim (2010).
9 J. Mráz, E. Gálová, H. Nohová, D. Vitková: Markers of exposure to Cyclohexanone, cyclohexane and cyclohexanol: 1,2- and 1,4-cyclohexanediol. Clin. Chem. 40, 1466–1468 (1994).
10 R. Fabre, R. Truhaut, M. Peron: Toxicological research on solvents to replace benzene. I. Study of cyclohexane. Arch. Mal. Prof. 13, 437–448 (1952).
11 P. L. Viola: On the presumed enzymatic dehydrogenation of cyclohexane to benzene. Boll. Soc. Ital. Biol. Sper. 36, 1960–1961 (1960).
12 J. Mráz, E. Gálová, H. Nohová, D. Vitková: 1,2- and 1,4-cyclohexanediol: major urinary metabolites and biomarkers of exposure to cyclohexane, cyclohexanone, and cyclohexanol in humans. Int. Arch. Occup. Environ. Health 71, 560–565 (1998).
13 D. Walter, A. Trabs, U. Knecht, H. J. Woitowitz: Toxikokinetische Daten zur Evaluierung eines BAT-Wertes für Cyclohexan. In: A. Rettenmeier, Ch. Feldhaus (eds.): Dokumentationsband über die 39. Jahrestagung der DGAUM. Rindt Fulda, 265–268 (1999).
14 H. Greim (ed.): Cyclohexane. MAK-Value Documentations. Volume 13, Wiley-VCH, Weinheim (1999).
15 C. Krause, M. Chutsch, M. Henke, M. Huber, C. Kliem, M. Leiske, W. Mailahn, C. Schulz, E. Schwarz, B. Seifert, D. Ullrich: Umwelt-Survey Band IIIc: Wohn-Innenraum. Raumluft. Wasser Boden Luft 4 (1991).
16 A. Perico, C. Cassinelli, F. Brugnone, P. Bavazzano, L. Perbellini: Biological monitoring of occupational exposure to cyclohexane by urinary 1,2- and 1,4-cyclohexanediol determination. Int. Arch. Occup. Environ. Health. 72, 115–120 (1999).
17 Bundesärztekammer: Richtlinie der Bundesärztekammer zur Qualitätssicherung quantitativer laboratoriumsmedizinischer Untersuchungen. Dt. Ärztebl. 105, A341–A355 (2008).
18 J. Angerer and G. Lehnert: Anforderungen an arbeitsmedizinisch-toxikologische Analysen – Stand der Technik. Dt. Ärztebl. 37, C1753–C1760 (1997).
19 J. Angerer, T. Göen, G. Lehnert: Mindestanforderungen an die Qualität von umweltmedizinisch-toxikologischen Analysen. Umweltmed. Forsch. Prax. 3, 307–312 (1998).

Author: *U. Knecht*
Examiners: *R. Heinrich-Ramm, K. H. Tieu*

11 Appendix

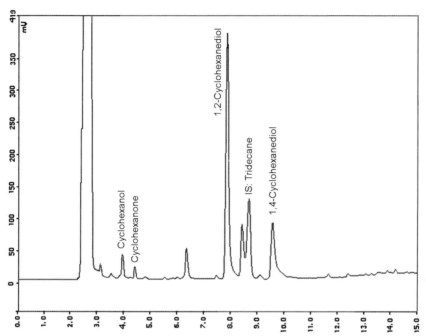

Fig. 2 Chromatogram of an urine sample of an individual exposed to cyclohexane (external exposure to 200 ml cyclohexane/m^3).

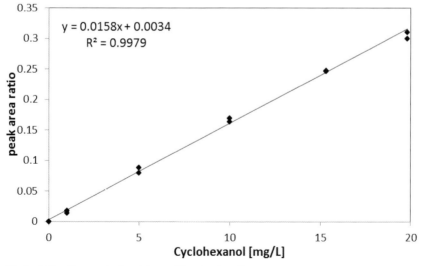

Fig. 3 Calibration graph of cyclohexanol in urine (n = 2).

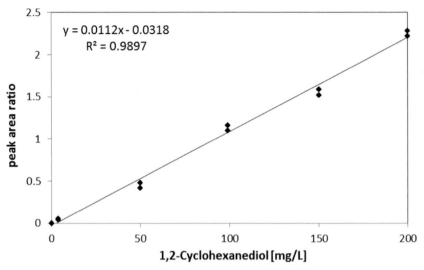

Fig. 4 Calibration graph of 1,2-cyclohexanediol in urine (n = 2).

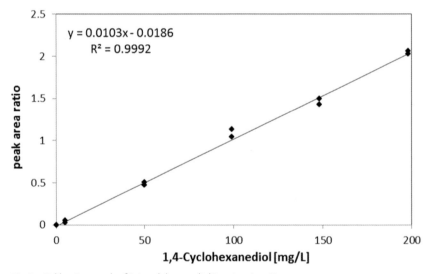

Fig. 5 Calibration graph of 1,4-cyclohexanediol in urine (n = 2).

N-(2,3-Dihydroxypropyl)-valine in blood as haemoglobin adduct of glycidol

Matrix:	Blood
Hazardous substances:	Glycidol, epichlorohydrin
Analytical principle:	Capillary gas chromatography with negative chemical ionisation and mass selective detection (GC-NCI-MS)
Completed in:	October 2008

Overview of parameters that can be determined with this method and the corresponding chemical hazardous substances:

Hazardous substances	CAS	Analyte	CAS
Glycidol	556-52-5	N-(2,3-Dihydroxypropyl)-valine	–
Epichlorohydrin	106-89-8		

Summary

N-(2,3-Dihydroxypropyl)-valine (DHPV) is the haemoglobin adduct of glycidol and the ultimate haemoglobin adduct of epichlorohydrin formed via different pathways.

To determine the adducts the erythrocytes are separated from whole blood and lysed. The globin is isolated from the haemoglobin solution by precipitation and the DHPV is derivatised with pentafluorophenyl isothiocyanate under addition of isotopically labelled N-(2,3-dihydroxypropyl)-$^{13}C_5$,^{15}N-valine-leucine anilide ($^{13}C_5$,^{15}N-DHPVLA) as internal standard (IS) and is cleaved-off from the residual globin chain by a modified Edman degradation. The derivatives of DHPV and of the IS are extracted with diethyl ether after addition of a saturated saline solution, washed several times and evaporated to dryness in a gentle stream of nitrogen. The free hydroxyl moieties of the alkylated valine are derivatised with acetone to form a ketal and separated from other components of the sample using capillary gas chromatography. Quantitative determination is carried out after negative chemical ionisation (NCI) using mass spectrometry with selected ion monitoring (SIM). Calibration is carried out using solutions of the dipeptide N-(2,3-dihydroxypropyl)-valine-leucine anilide (DHPVLA) added to of a pooled human globin in formamide and processed analogously to the samples.

N-(2,3-Dihydroxypropyl)-valine (DHPV)

Within day precision:	Standard deviation (rel.)	s_w = 6.9% or 4.6%
	Prognostic range	u = 17.7% or 11.9%
	at a spiked concentration of 50 or 450 pmol DHPV per g globin and where n = 6 determinations	
Day to day precision:	Standard deviation (rel.)	s_w = 6.0% or 10.1%
	Prognostic range	u = 15.4% or 32.1%
	at a spiked concentration of 40 or 100 pmol DHPV per g globin and where n = 5 or 4 determinations	
Accuracy:	Recovery rate	r = 98% or 102%
	at a spiked concentration of 50 or 450 pmol DHPV per g globin and where n = 8 or 6 determinations	
Detection limit:	10 pmol DHPV per gram globin	
Quantitation limit:	25 pmol DHPV per gram globin	

Glycidol (2,3-epoxy-1-propanol)

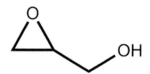

Glycidol is a racemate of two stereoisomers (right and left enantiomer). As a bifunctional alkylating agent it is an important starting compound for the synthesis of glycidyl and glyceryl derivatives, which are in turn used in the production of surfactants, plastics, paints, photochemicals, pharmaceuticals and biocides. Glycidol acts as a stabiliser for plastic polymers, as an additive for oils and synthetic hydraulic fluids and as a solvent for several epoxy resins. At room temperature, glycidol is a low-viscous, colourless and odourless liquid (molecular weight: 74.08 g/mol; boiling point: 166.1°C; vapour pressure at 25°C: 1.2 hPa). It is miscible with water, alcohols, ketones, esters, ethers and aromatic compounds in any ratio. On the other hand, glycidol is practically insoluble in aliphatic hydrocarbons. Glycidol reacts vigorously with strong oxidants and nitrates [1–3]. Glycidol is not known to occur naturally. There are no current data on the world's annual production available, older sources describe an annual production of 5,000 tons in the USA and about 4,900 potentially exposed individuals [2, 3].

A comprehensive presentation of the toxicological profile of glycidol is published in the MAK documentation of the *Deutsche Forschungsgemeinschaft* (DFG), as well as in monographs of the IARC (International Agency for Research on Cancer) and of the NTP (National Toxicology Program) [1–3]. Glycidol is classified by the Commission as carcinogen of Category 2 [4]. Other national and international committees agree with this classification [2, 3]. The MAK value of 50 mL/m^3 established in 1958 on the basis of the former TLV value was therefore withdrawn in 2000. Glycidol can be absorbed through the skin and is therefore designated with an "H" (risk of skin absorption). Due to lack of available data, the substance is not designated with "S" (skin sensitisation hazard).

Investigations on the metabolism in rats showed that 87–92% of an orally administered glycidol dose is absorbed by the gastrointestinal tract. 72 hours after administration, 7–8% of that dose was still detectable in the tissues [2].

After intraperitoneal injection, the metabolites S-(2,3-dihydroxypropyl)-glutathione, S-(2,3-dihydroxypropyl)-cysteine and β-chlorolactic acid are found in the urine [3]. Apart from the direct alkylation of nucleophilic molecules, a conversion of glycidol to 3-chloropropane-1,2-diol (α-chlorohydrin) by direct reaction with the hydrochloric acid in the stomach presumably takes place. The 3-chloropropane-1,2-diol formed can both be conjugated with L-glutathione catalysed by glutathione S-transferase and oxidised enzymatically to β-chlorolactic acid. Hydrolysis of glycidol to glycerol catalysed by epoxide hydrolases could be demonstrated by incubation with lung and rat liver microsomes [1].

As a bifunctional alkylating agent glycidol is capable of reacting with relevant physiological macromolecules. With calf thymus DNA it forms the DNA adducts 3-(2,3-dihydroxypropyl)-dUrd and 3-(2,3-dihydroxypropyl)-dThd [1].

In addition, glycidol reacts directly with the terminal valine of the human haemoglobin chains forming the haemoglobin adduct N-(2,3-dihydroxypropyl)-valine (DHPV) [5]. In our own investigations aliquots of a non-smoker's haemolysate were incubated with increasing concentrations of racemic glycidol (25–500 µM). A dose-dependent linear increase in the DHPV adduct level was found. Thus, the binding kinetics of glycidol in human haemoglobin could be demonstrated for the first time *in vitro* [6]. In addition, DHPV is the ultimate haemoglobin adduct of epichlorohydrin formed via a number of postulated pathways [5]. DHPV formation after reductive conversion of rat haemolysate could additionally serve as indirect evidence for the fact that glycerine aldehyde and glycid aldehyde form haemoglobin adducts [7].

With regard to DHPV adduct levels in humans, studies found mean values of more than 10 pmol DHPV/g globin (21.1 ± 17.1 pmol DHPV/g globin (n = 7) or 13.1 ± 12.4 pmol DHPV/g globin (n = 8)) in smoker's globin in German collectives. The mean values of non-smokers were just below 10 pmol DHPV/g globin (7.3 ± 2.7 pmol DHPV/g globin (n = 8) or 6.8 ± 3.2 pmol DHPV/g globin (n = 3)) [8]. On the other hand, in a Swedish collective the DHPV adduct levels for non-smokers and smokers were determined to be 2.1 ± 1.1 pmol DHPV/g globin (n = 6) and 9.5 ± 2.2 pmol DHPV/g globin (n = 4), respectively [8].

Recent investigations in three collectives of disaster relief forces (n_{total} = 554) potentially exposed to epichlorohydrin released in a hazardous substance accident, showed that the primary haemoglobin adduct N-(3-chloro-2-hydroxypropyl)-valine (CHPV) could be detected in 6 persons with an adduct level of 30–80 pmol CHPV/g globin (quantitation limit: 20 pmol/g globin) [11]. On the other hand, using a method with a comparable quantitation limit of 25 pmol/g globin, DHPV as the ultimate haemoglobin adduct of epichlorohydrin could not be determined in the same samples [11]. Based on the *in vitro* investigations of binding kinetics, it can be assumed that the biomarker DHPV is generally applicable as biomarker of exposure to glycidol. As biomarker of exposure to epichlorohydrin, the determination of the p4rimary haemoglobin adduct CHPV is to be preferred at the present state of the art.

Epichlorohydrin

Am comprehensive review on epichlorohydrin is available as part of the method "N-(3-chloro-2-hydroxypropyl)-valine as haemoglobin adduct of epichlorohydrin". This method is also published in "The MAK-Collection. Part IV: Biomonitoring Methods. Volume 13".

Author: *M. Müller*
Examiners: *Th. Göen, E. Eckert, Th. Schettgen*

N-(2,3-Dihydroxypropyl)-valine in blood as haemoglobin adduct of glycidol

Matrix:	Blood
Hazardous substances:	Glycidol, epichlorohydrin
Analytical principle:	Capillary gas chromatography with negative chemical ionisation and mass selective detection (GC-NCI-MS)
Completed in:	October 2008

Contents

1	General principles
2	Equipment, chemicals and solutions
2.1	Equipment
2.2	Chemicals
2.3	Solutions
2.4	Internal standard
2.5	Calibration standards
3	Specimen collection and sample preparation
3.1	Preparation of the erythrocyte lysate
3.2	Isolation of the globin
3.3	Derivatisation of the globin
4	Operational parameters
4.1	Operational parameters for gas chromatography
4.2	Operational parameters for mass spectrometry
5	Analytical determination
6	Calibration
7	Calculation of the analytical result
8	Standardisation and quality control
9	Evaluation of the method
9.1	Precision
9.2	Accuracy
9.3	Detection limit and quantitation limit

9.4 Sources of error
10 Discussion of the method
11 References
12 Appendix

1 General principles

N-(2,3-Dihydroxypropyl)-valine (DHPV) is the haemoglobin adduct of glycidol [6] and the ultimate haemoglobin adduct of epichlorohydrin formed via different pathways [5] (Figure 1).

To determine the adducts, erythrocytes are separated from the whole blood and lysed. The globin is isolated from the haemoglobin solution by precipitation and DHPV is derivatised with pentafluorophenyl isothiocyanate under addition of isotopically labelled N-(2,3-dihydroxypropyl)-$^{13}C_5,^{15}N$-valine-leucine anilide ($^{13}C_5,^{15}N$-DHPVLA) [12] as internal standard (IS) and is cleaved by a modified Edman degradation. After addition of a saturated saline solution, the derivatives of DHPV and the IS are extracted with diethyl ether, washed several times and evaporated to dryness under a gentle stream of nitrogen. The free hydroxy moieties of the alkylated valine are then derivatised with acetone to form a ketal according to Paulsson et al. [13] (see Figure 2) and separated from the other components of the sample using capillary gas chromatography. Quantitative determination is carried out after negative chemical ionisation by mass spectrometry in the selected ion monitoring mode (SIM) using the ion traces $m/z = 410$ as well as $m/z = 390$ for the analyte DHPV and $m/z = 415$ as well as $m/z = 395$ for the internal standard $^{13}C_5,^{15}N$-DHPV. Calibration is carried out using solutions of the dipeptide N-(2,3-dihydroxy-propyl)-valine-leucine anilide (DHPVLA) [12] added to a pooled human globin in formamide and processed analogously to the protein samples.

2 Equipment, chemicals and solutions

2.1 Equipment

- Gas chromatograph/mass spectrometer with split/splitless injector, negative chemical ionisation (NCI) unit, and data processing system
- Gas chromatographic column: length: 30 m; inner diameter: 0.25 mm; stationary phase: (50%-phenyl)-methylpolysiloxane; film thickness: 0.15 µm (e.g. DB-17HT, J&W Scientific, No. 122-1831)
- Analytical balance (e.g. Sartorius)
- Roll mixer (e.g. Karl Hecht KG)
- Horizontal shaker (e.g. Janke & Kunkel)

- Laboratory shaker (e.g. Janke & Kunkel)
- Lyophilisation unit (e.g. Steris)
- Ultrasonic bath (e.g. Bandelin electronic)
- Drying oven (e.g. Heraeus)
- Evaporation unit (e.g. Gebr. Liebisch GmbH & Co.)
- EDTA monovettes (e.g. Sarstedt)
- 7 mL and 20 mL Glass tubes with screw caps (e.g. Schott)
- 10 µL Syringe for gas chromatography (e.g. Hamilton)
- Microliter pipettes, with adjustable volumes between 2 and 20 µL, 10 and 100 µL as well as between 100 and 1000 µL (e.g. Eppendorf)
- Finn pipette 1–5 mL
- 1 mL Glass pipettes
- Laboratory centrifuge, at least 3500 g (e.g. Hettich)
- 2 mL Glass vials (e.g. Agilent)
- 100 µL Microinserts (e.g. Agilent)
- 10, 25, 100 and 1000 mL Volumetric flasks
- 50 mL and 250 mL Glass beakers

2.2 Chemicals

- 2-Propanol p.a. (e.g. Merck, No. 1.09634)
- Ethyl acetate p.a. (e.g. Merck, No. 1.09623)
- n-Hexane p.a. (e.g. Merck, No. 1.07023)
- Sodium chloride p.a. (e.g. Merck, No. 1.06404)
- Sodium carbonate, anhydrous p.a. (e.g. Merck, No. 1.06392)
- Sodium hydrogen carbonate, p.a. (e.g. Merck, No. 1.06329)
- Hydrochloric acid 37% (e.g. Merck, No. 1.00317)
- Sodium hydroxide pellets p.a. (e.g. Merck, No. 1.06498)
- Highly purified water or bidist. water
- Diethyl ether p.a (e.g. Merck, No. 1.00921)
- Formamide ultrapure (e.g. Amersham Life Science, No. US75828)
- Ethanol p.a. (e.g. Merck, No. 1.00983)
- N-(2,3-Dihydroxypropyl)-valine-leucine anilide (DHPVLA, e.g. Kadem Custom Chem GmbH)
- N-(2,3-Dihydroxypropyl)-$^{13}C_5$,^{15}N-valine-leucine anilide ($^{13}C_5$,^{15}N-DHPVLA, e.g. Kadem Custom Chem GmbH)
- Pentafluorophenyl isothiocyanate for GC, ≥97% (e.g. Fluka, No. 76755)
- Toluene p.a. (e.g. Fluka, No. 34938)
- Acetone p.a. (e.g. Merck, No. 1.07021)
- Sulfuric acid 96% suprapure (e.g. Merck, No. 1.00714)
- Nitrogen 5.0 (e.g. Linde)
- Helium 5.0 (e.g. Linde)
- Methane 5.5 (e.g. Linde)

Preparation of standard globin:
To obtain standard globin, 2 mL haemolysate each are prepared according to Section 3 from whole blood samples of two non-smokers (female, male). The isolated freeze-dried globins are then thoroughly mixed and stored at −20°C. Under these conditions the globin samples can be kept up to one year.

2.3 Solutions

- 50 mM Hydrochloric acid in 2-propanol
  4.1 mL 37% hydrochloric acid are pipetted into a 1000 mL volumetric flask containing about 500 mL 2-propanol. The flask is then made up to the mark with 2-propanol.
- 0.9% Saline solution
  9 g sodium chloride are weighed into a 1000 mL volumetric flask and dissolved in bidist. water. The flask is then made up to the mark with bidist. water.
- Saturated saline solution
  10 mL bidist. water are placed in a 50 mL glass beaker. Sodium chloride is carefully added under light shaking until a sediment of undissolved sodium chloride (approx. 3.6 g NaCl) remains at the bottom of the beaker.
- 1 M Sodium hydroxide solution
  4 g sodium hydroxide are weighed into a 250 mL glass beaker and dissolved in bidist. water. The solution is transferred to a 100 mL volumetric flask which is then made up to the mark with bidist. water.
- 0.1 M Sodium carbonate solution
  1.06 g sodium carbonate are weighed into a 100 mL volumetric flask and dissolved in bidist. water. The flask is then made up to the mark with bidist. water.
- 0.1 M Sodium hydrogen carbonate solution
  0.84 g sodium hydrogen carbonate are weighed into a 100 mL volumetric flask. The flask is then made up to the mark with bidist. water.

The solutions can be stored in the refrigerator at 4°C for at least one month.

- Acidic acetone (reagent for ketal derivatisation)
  About 5 mL acetone are placed in a 10 mL volumetric flask and 0.1 mL 96% sulfuric acid are added using a glass pipette. The flask is then made up to the mark with acetone.

The reagent is to be prepared freshly each time just before use.

2.4 Internal standard

- Stock solution (1 mM $^{13}C_5,^{15}N$-DHPV)
  9.6 mg N-(2,3-dihydroxypropyl)-$^{13}C_5,^{15}N$-valine-leucine anilide ($^{13}C_5,^{15}N$-

DHPVLA) are weighed exactly into a 25 mL volumetric flask which is then made up to the mark with ethanol.
- Working solution (1 µM $^{13}C_5,^{15}N$-DHPV)
  10 µL of the stock solution are pipetted into a 10 mL volumetric flask containing 5 mL ethanol. The volumetric flask is then made up to the mark with ethanol.

The solutions can be stored in the refrigerator at 4°C for at least 6 months without noteworthy losses.

2.5 Calibration standards

- Stock solution (1 mM DHPV)
  9.5 mg N-(2,3-dihydroxypropyl)-valine-leucine anilide (DHPVLA) are weighed exactly into a 25 mL volumetric flask which is then made up to the mark with ethanol.
- Working solution (1 µM DHPV)
  10 µL of the stock solution are pipetted into a 10 mL volumetric flask containing 5 mL ethanol. This volumetric flask is then made up to the mark with ethanol.

The solutions can be stored in the refrigerator at 4°C for at least 6 months without noteworthy losses.

To prepare the calibration standards, 100 mg standard globin each are weighed in seven 7 mL screw cap glass tubes and are dissolved in 3 mL formamide each. Subsequently, the solutions are spiked with the working solutions of the analyte (DHPV) and of the internal standard ($^{13}C_5,^{15}N$-DHPV) according to Table 1. The calibration standards are processed according to Section 3.3 in the same way as the samples and then analysed according to Section 4 and Section 5.

The calibration standards must be prepared anew for each analytical series.

Table 1 Pipetting scheme for the preparation of the calibration standard solutions.

Calibration standard	DHPV (1 µM working solution) [µL]	IS $^{13}C_5,^{15}N$-DHPV (1 µM working solution) [µL]	DHPV in the sample: analyte/IS [pmol/g globin]
1	2	100	20/1000
2	5	100	50/1000
3	10	100	100/1000
4	20	100	200/1000
5	50	100	500/1000
6	100	100	1000/1000
7 (only IS)	0	100	0/1000

3 Specimen collection and sample preparation

3.1 Preparation of the erythrocyte lysate

Blood samples (approx. 9 mL) are collected with an EDTA monovette. The blood sample thus obtained is centrifuged at 800 g for 10 min in order to separate the erythrocytes from the blood plasma. The supernatant plasma is removed with a glass pipette and discarded. 5 mL 0.9% saline solution are added to the remaining erythrocyte fraction, mixed thoroughly and are again centrifuged at 800 g for 10 min. The supernatant is removed again using a glass pipette and discarded. This washing process is repeated until the supernatant is clear and colourless (three times for fresh blood samples as a rule). For haemolysis, 4 mL highly purified water are added to the erythrocyte fraction and the solution is deep frozen at −20°C for at least 60 min.

Our own investigations have shown that erythrocyte lysates which are deep frozen at −20°C and treated with racemic glycidol can be stored up to two years without noteworthy losses of DHPV adducts. Deep frozen erythrocyte lysates should be transported on dry ice.

3.2 Isolation of the globin

The erythrocyte lysate is thawed, equilibrated to room temperature and homogenised by careful swirling. To isolate 200–250 mg globin, 2 mL of the haemolysate is pipetted into a 20 mL screw cap glass tube containing 12 mL 50 mM hydrochloric acid in 2-propanol. The sample is mixed for 15 min on a roll mixer (20 U/min) and then centrifuged at 3500 g for 10 min. The supernatant is decanted off into a new 20 mL screw cap glass tube and 8 mL ethyl acetate are slowly added. Cooling to 4°C for 90 min (refrigerator) completes precipitation of the globin. The globin is centrifuged off (800 g, 10 min), and the supernatant is discarded. The residue is resuspended in 10 mL ethyl acetate, centrifuged again (800 g, 10 min) and the supernatant is discarded again. This washing process is repeated until the supernatant is clear and colourless (approx. three times when preparing fresh blood samples as a rule). The globin is finally resuspended in 5 mL n-hexane and centrifuged at 800 g for 10 min. The supernatant is discarded and the globin isolate is dry frozen. Our own investigations have shown that globin samples stored at −20°C for up to two years exhibit no noteworthy changes in DHPV adduct levels produced *in vitro* with racemic glycidol.

3.3 Derivatisation of the globin

For analysis, 100 mg globin are weighed into a 7 mL screw cap glass tube to which 3 mL formamide, 100 µL of the working solution of the IS (1 µM $^{13}C_5,^{15}N$-DHPV) and 30 µL 1 M sodium hydroxide solution are added. The standards for calibration are prepared as described under Section 2.5 and 30 µL of 1 M sodium hydroxide solution are added to each of them.

In order to ensure a homogeneous solution, the sealed screw cap glass tube is placed in an ultrasonic bath for 30 min. For derivatisation, 15 µL pentafluorophenyl isothiocyanate are then added. The sample solution is mixed overnight on a roll mixer (20 U/min) at room temperature and then incubated on a roll mixer (20 U/min) at 45°C for 120 min in a drying oven. After equilibration to room temperature, 400 µL saturated saline solution and 4 mL diethyl ether are added, the solutions are thoroughly mixed using a laboratory shaker for 60 s and then centrifuged at 3500 g for 5 min. The ether phase is carefully removed and transferred to a new 7 mL screw cap glass tube. 3 mL diethyl ether are added to the remaining aqueous phase, thoroughly mixed on a laboratory shaker for 60 s and then centrifuged at 3500 g for 5 min. The ether phase is carefully removed and combined with the first ether phase. The combined ether phases are evaporated to dryness under a stream of nitrogen and dissolved in 1.5 mL toluene. The toluene phase is first washed with 2 mL highly purified water and then with 2 mL 0.1 M sodium carbonate solution (60 s on the laboratory shaker each time). Each washing process is followed by centrifugation at 3500 g for 5 min. After each washing step the toluene phase is transferred into a new 7 mL screw cap glass tube. After the final washing step the toluene phase is transferred to a new 7 mL screw cap glass tube and then evaporated to dryness under a gentle stream of nitrogen using moderate heating at about 30°C. A glassy amber residue should remain on the bottom of the glass; if necessary, residues on the walls of the vial are dissolved in toluene (100–200 µL) and dried again. Subsequently, the residue is dissolved in 100 µL freshly prepared acidic acetone. The sample tube is sealed and the solution is mixed overnight on a horizontal shaker at 30 U/min. Then, 150 µL 0.1 M sodium hydrogen carbonate solution are added to the sample and mixed thoroughly on a laboratory shaker for 60 s. This is followed by the addition of 1 mL toluene, mixing on a laboratory shaker for 60 s as well as centrifugation at 3500 g for 5 min. The toluene phase is transferred to a new 7 mL screw cap glass tube and washed two times with 1 mL highly purified water (mixing on a laboratory shaker for 60 s each). Each washing process is followed by centrifugation at 3500 g for 5 min, after which the toluene phase is transferred to a new 7 mL screw cap glass tube. After the final washing process the toluene phase is transferred to a 2 mL glass vial and evaporated to dryness under a gentle stream of nitrogen while heating the vial to 30°C. A glassy amber residue should stick to the bottom of the glass; if necessary, residues on the walls of the vial are dissolved in toluene (100–200 µL) and dried again. The residue is then dissolved in 30 µL toluene and transferred to a 100 µL micro-insert.

If the analytical series is not immediately continued, the 100 µL micro-insert can be placed in a 2 mL glass vial. The sealed sample can thus be stored in the deep freezer at −80°C for at least 1 month without noteworthy losses.

Before analysis, the micro-insert is held between the fingertips and the toluene is removed under a gentle stream of nitrogen. Subsequently, the 100 µL micro-insert is placed in a 2 mL glass vial. Then the residue is dissolved in 10 µL toluene by injecting toluene into the micro-insert with a 10 µL-syringe for gas chromatography. The residue is dissolved by repeated careful drawing up with a syringe. 1 µL of the solution is injected for gas chromatographic/mass spectrometric analysis.

4 Operational parameters

4.1 Operational parameters for gas chromatography

Capillary column:	Material:	Quartz glass (fused silica)
	Stationary phase:	DB-17HT
	Length:	30 m
	Inner diameter:	0.25 mm
	Film thickness:	0.15 µm
Temperature:	Column:	1 min at 90°C; then increase at 25°C/min to 120°C; then increase at 10°C/min to 240°C; then increase at 25°C/min to 310°C; 10 min at final temperature
	Injector:	280°C
	Transfer line:	280°C
Carrier gas:	Helium 5.0 with a column pre-pressure of 12.2 psi, constant flow: 1.2 mL/min	
Sample volume:	1 µL (pulsed splitless, pulse time: 90 s)	

4.2 Operational parameters for mass spectrometry

Ionisation type:	Negative chemical ionisation (NCI)
Source temperature:	150°C
Quadrupol temperature:	106°C
Reagent gas:	Methane 5.5 (pre-pressure 1.5 bar; mass flow counter: 45%; MSD-vacuum: ~2.0×10^{-4} Torr)
Detection:	Selected ion monitoring (SIM)
Measuring time per ion:	50 ms

All other parameters must be optimised in accordance with the manufacturer's instructions.

Figure 3 shows the ion chromatogram of a calibration standard.

5 Analytical determination

1 µL of the prepared sample is injected into the GC-NCI-MS system without gas flow division (pulsed splitless).

The Edman derivatives of the unlabelled N-(2,3-dihydroxypropyl)-valine (DHPV) and of the labelled internal standard ($^{13}C_5, ^{15}N$-DHPV) are present in the form of a diastereomeric pair on account of the chiral centre at the C-2 of the hydroxypropyl side chain and are eluting as two baseline-separated peaks of about the same signal intensity. Identification is carried out using the retention times and the characteristic ion traces (Table 2). As regards the retention times of DHPV, minimal differences were found between the calibration standard and the samples obtained from incubations of erythrocyte lysate with racemic glycidol, for which presumably deviations in the mixing ratio between the stereoisomers are responsible (Figure 3 and Figure 4). For quantitative evaluation the corresponding peaks of the ion traces are integrated and the peak areas of both peaks thus obtained are added.

The retention times given in Table 2 are intended to be a rough guide only.

Table 2 Retention times of the analytes as well as the analysed and evaluated ion traces.

Analyte	Retention time [min]	Analysed ion traces [m/z]	Evaluated ion trace [m/z]
DHPV (peak1, peak2) Calibration	13.15; 13.35	410 [M-CO]$^-$ 390 [M-CO-HF]$^-$	410
DHPV (peak1, peak2) Incubation	13.20; 13.40	410 [M-CO]$^-$ 390 [M-CO-HF]$^-$	410
$^{13}C_5, ^{15}N$-DHPV (peak1, peak2) Internal standard	13.15, 13.35	415 [M-*CO]$^-$ 395 [M-*CO-HF]$^-$	415

* ^{13}C-labelled

6 Calibration

Calibration graphs are obtained by plotting the quotients of the peak areas of the analyte and the internal standard against the spiked concentration in pmol/g globin (Figure 5). The calibration graph is linear between 20 and 1000 pmol DHPV/g globin.

7 Calculation of the analytical result

To determine the analyte concentration in a sample, the peak area of the analyte is divided by the peak area of the internal standard. The quotient thus obtained is inserted into the corresponding calibration graph (see Section 6). The DHPV concentration level of the sample is obtained in pmol/g globin. Use the following algorithm to convert into the unit µg/L blood (c = concentration):

$$c \, [\mu g/L \text{ blood}] = c \, [\text{pmol/g globin}] \times (191 \times 10^{-6} \, \mu g/\text{pmol}) \times 144 \, g/L.$$

8 Standardisation and quality control

Quality control of the analytical results is carried out as stipulated in the guidelines of the *Bundesärztekammer* (German Medical Association) [14]. To check precision, a globin control sample with a known concentration level of DHPV is analysed within each analytical series. As a material for quality control is not commercially available, it must be prepared in the laboratory by incubation of the erythrocyte lysate of a non-smoker with racemic glycidol.

The globin is isolated from the control material and kept deep frozen at –20°C. Stored in this manner control material can be kept for at least 2 years. The nominal value and the tolerance ranges of the quality control material are determined in a pre-analytical period (one analysis of the control material on each of 20 different days) [15, 16].

External quality control can be realised by taking part in external quality assessment schemes. According to our present knowledge, the round-robins carried out in Germany for occupational-medical and environmental-medical toxicological analyses do not, however, include the analysis of DHPV in their external quality control program [17].

9 Evaluation of the method

9.1 Precision

To determine the within day precision six calibration standards with a DHPV concentration of 50 or 450 pmol/g globin were prepared and analysed in a row. Determination of day to day precision was carried out by preparing and analysing calibration standards with a DHPV concentration of 40 and 100 pmol/g globin on five and four different days, respectively. The results are listed in Table 3.

Table 3 Within day and day to day precision for the determination of DHPV.

Spiked DHPV level [pmol/g]	Number of determinations	Relative standard deviation [%]	Prognostic range [%]
Within day precision			
50	6	6.9	17.7
450	6	4.6	11.9
Day to day precision			
40	5	6.0	15.4
100	4	10.1	32.1

9.2 Accuracy

Determination of the accuracy of the method was carried out using pooled globin at a spiked concentration of 50 and 450 pmol/g globin. The obtained recovery rates are given in Table 4.

Table 4 Relative recovery rates for DHPV in globin.

Spiked DHPV level [pmol/g]	Number of determinations	Mean relative recovery [%]	Range [%]
50	8	98	88–104
450	6	102	96–107

9.3 Detection limit and quantitation limit

A detection limit of 10 pmol DHPV/g globin was estimated for the procedure based on a signal to noise ratio of 3 : 1. The quantitation limit was estimated to be 25 pmol DHPV per g globin.

9.4 Sources of error

Care must be taken that a glassy amber residue is present on the bottom of the sample vial after evaporation of the toluene phase. A white gritty sediment indicates considerable analyte losses. A new preparation of the sample with diethyl ether from a freshly opened flask has produced the desired result in all cases.

On account of the highly concentrated sample solution, contamination of the injector is easily possible. In the event of interference peaks, a change of the liner and if necessary of the gold seal and a cut-off of the column is recommended.

10 Discussion of the method

A new method to detect DHPV has been developed based on standard procedures of the *Deutsche Forschungsgemeinschaft* (German Research Foundation) for the detection of haemoglobin adducts [18]. The DHPV is isolated according to the modified Edman degradation and is derivatised in a final step with acetone to a ketal [13]. After gas chromatographic separation and electron impact ionisation (EI) the derivative thus obtained can be determined with a detection limit of 25 pmol DHPV/g globin. By using negative chemical ionisation (NCI) and reducing the sample volume by evaporation to a final volume of 10 µL, a detection limit of 10 pmol DHPV/g globin is obtained.

For the new procedure the usual valine-leucine-dipeptide adduct standards [18] were developed further. By the introduction of stable isotopes ($^{13}C_5, ^{15}N$) in the valine a synthesis platform for any selected adduct standard for alkylating substances is obtained [12]. In the present case, the internal standard substance $^{13}C_5, ^{15}N$-DHPVLA has been synthesised. In this way, the principle of isotope dilution for valine leucine dipeptide adduct standards can be used for the first time thus making the previous internal standardisation using N^2-ethoxyethyl-valine-alanine anilide no longer necessary. The method of isotope dilution improves sensitivity and reproducibility in the routine analysis considerably and reduces the susceptibility for sources of error.

To check the applicability of the method in practice a differentiation between smoker's and non-smoker's globin was originally planned. In the literature [8] mean DHPV values of smoker's globin in German collectives of >10 pmol /g globin (21.1 ± 17.1 pmol DHPV/g globin (n = 7) or 13.1 ± 12.4 pmol DHPV/g globin (n = 8)) were found; the mean DHPV levels of non-smokers were just below 10 pmol DHPV/g globin (7.3 ± 2.7 pmol DHPV/g globin (n = 8) or 6.8 ± 3.2 pmol DHPV/g globin (n = 3)). However, our own investigations in three globin samples of smokers (cigarette consumption: 10–40 per day) revealed no detectable DHPV levels (detection limit: 10 pmol DHPV/g globin).

In a study by Hindsø Landin et al. [8] the formation of DHPV in globin of smokers was associated with the uptake of the epichlorohydrin metabolites glycidol and 3-chloropropane-1,2-diol from cigarette smoke [9]. However, the current list of additives for tobacco products does not contain these substances [10]. Though glycerol, a potential precursor of glycidol, is still used by a number of manufacturers, it only plays a minor role as additive compared with 1,2-propylene glycol. Due to its different chemical structure, 1,2-propylene glycol forms no DHPV adducts with haemoglobin. The lack of exposure from cigarette smoke may serve as explanation for our negative findings and the wide variations in DHPV levels in the early investigations (due to differing exposures to additives depending on the brand smoked).

However, to demonstrate the suitability of the analytical procedure for detecting DHPV in haemoglobin, *in vitro*-investigations on haemoglobin adduct formation were carried out. For this purpose aliquots of a non-smoker haemolysate were incubated with increasing concentrations of racemic glycidol (25–500 µM). As expected for a direct-acting alkylating agent, a dose-dependent linear increase in DHPV adduct levels was found (Figure 6). Thus, the binding kinetics of glycidol in human haemoglobin could be demonstrated *in vitro* for the first time [6].

Hence, the applicability of the new procedure to detect DHPV in modified haemoglobin could be confirmed. The sensitivity of the method can be considered as sufficient to demonstrate exposure. Thus, in studies with rats an adduct level of 44 pmol DHPV/g globin was still detectable 30 days after single i.p. injection of 432 µmol epichlorohydrin/kg body weight, which corresponds to an adduct level of 88 pmol DHPV/g globin immediately after exposure [5].

11 References

1. H. Greim (ed.): Glycidol. The MAK-Collection. Part I: MAK Value Documentations. Volume 20. Wiley-VCH Verlag, Weinheim (2003).
2. International Agency for Research on Cancer (IARC): Some Industrial Chemicals: Glycidol. Monographs on the Evaluation of Carcinogenic Risks to Humans. Vol. 77, 469–486 (2000).
3. National Toxicology Program (NTP): Report on Carcinogens. 11th edition. U.S. Department of Health and Human Services, Public Health Service, National Toxicology Program, Research Triangle Park, NC. 130–131 (2004).
4. Deutsche Forschungsgemeinschaft (DFG): List of MAK and BAT values 2012. 48th issue, Wiley-VCH-Verlag, Weinheim (2012).
5. H. Hindsø Landin, S. Osterman-Golkar, V. Zorcec, M. Törnqvist: Biomonitoring of epichlorohydrin by hemoglobin adducts. Anal. Biochem. 240, 1–6 (1996).
6. M. Müller, V. Belov, A. de Meijere, J. Bünger, B. Emmert, A. Heutelbeck, E. Hallier: Entwicklung eines neuen Biomonitoringverfahrens zur Bestimmung des Hämoglobinadduktes N-(2,3-Dihydroxypropyl)-Valin nach einer Epichlorhydrin-exposition. Verh. Dt. Ges. Arbeitsmed. 45, 531–533 (2005).
7. H. Hindsø Landin, E. Tareke, P. Rydberg, U. Olsson, M. Törnqvist: Heating of food and haemoglobin adducts from carcinogens: possible precursor role of glycidol. Food Chem. Toxicol. 38, 963–969 (2000).
8. H. Hindsø Landin, T. Grummt, C. Laurent, A. Tates: Monitoring of occupational exposure to epichlorohydrin by genetic effects and hemoglobin adducts. Mutat. Res. 381, 217–226 (1997).
9. J.N. Schumacher, C.R. Green, F.W. Bes, M.P. Newell: Smoke composition. An extensive investigation of the water-soluble portion of cigarette smoke. J. Agric. Food Chem. 25, 310–320 (1977).
10. Bundesministerium für Ernährung, Landwirtschaft und Verbraucherschutz (BMELV). Tabakzusatzstoff-Datenbank. Verbraucherschutz und Informationsrechte. Gesundheit. Nichtraucherschutz, URL: http://www.bmelv.de/cln_181/SharedDocs/Standardartikel/Verbraucherschutz/Gesundheitsmarkt/NichtRauchen/Tabakzusatzstoffe.html; visited on 28.09.2010.
11. K.M. Wollin, M. Bader, M. Müller, W. Lilienblum, M. Csicsaky: Assessment of long-term health effects by means of haemoglobin adducts of 1-chloro-2,3-epoxypropane (ECH) after

accidental exposure. Naunyn-Schmiedeberg's Arch. Pharmacol. 377, Supplement 1, 92 (2008).

12 V.N. Belov, M. Müller, O. Ignatenko, E. Hallier, A. de Meijère: Facile access to isotopically labeled valylleucyl anilides as biomarkers for the quantification of hemoglobin adducts to toxic electrophiles. Eur. J. Org. Chem. 23, 5094–5099 (2005).

13 B. Paulsson, I. Athanassiadis, P. Rydberg, M. Törnqvist: Hemoglobin adducts from glycidamide: acetonization of hydrophilic groups for reproducible gas chromatography/tandem mass spectrometric analysis. Rapid Commun. Mass Spectrom. 17, 1859–1865 (2003).

14 Bundesärztekammer: Richtlinie der Bundesärztekammer zur Qualitätssicherung quantitativer laboratoriumsmedizinischer Untersuchungen. Dt. Ärztebl. 100, A3335–A3338 (2003).

15 J. Angerer and G. Lehnert: Anforderungen an arbeitsmedizinisch-toxikologische Analysen – Stand der Technik. Dt. Ärztebl. 37, C1753–C1760 (1997).

16 J. Angerer, T. Göen, G. Lehnert: Mindestanforderungen an die Qualität von umweltmedizinisch-toxikologischen Analysen. Umweltmed. Forsch. Prax. 3, 307–312 (1998).

17 Ringversuch Nr. 34. Qualitätsmanagement in der Arbeits- und Umweltmedizin, Projektgruppe Qualitätssicherung, Organisation: Institut für Arbeits-, Sozial- und Umweltmedizin der Universität Erlangen-Nürnberg (2004).

18 N.J. Van Sittert: N2-Cyanoethyl-Valin, N2-Hydroxyethyl-Valin, N-Methyl-Valin (zum Nachweis einer Belastung/Beanspruchung durch Acrylnitril, Ethylenoxid sowie methylierende Substanzen) in: Deutsche Forschungsgemeinschaft (DFG): Analytische Methoden zur Prüfung gesundheitsschädlicher Arbeitsstoffe: Analysen in biologischem Material; Hrsg. Greim H.; Wiley-VCH Verlag GmbH, Weinheim, 12th issue (1996).

Author: *M. Müller*
Examiners: *Th. Göen, E. Eckert, Th. Schettgen*

11 Appendix

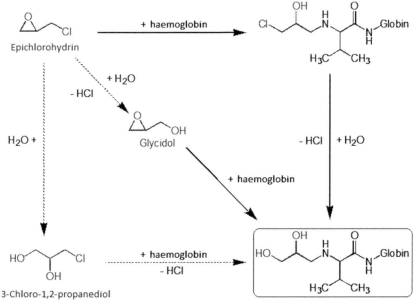

Fig. 1 Proposed reaction pathways of epichlorohydrin with human haemoglobin (modified according to [5]).

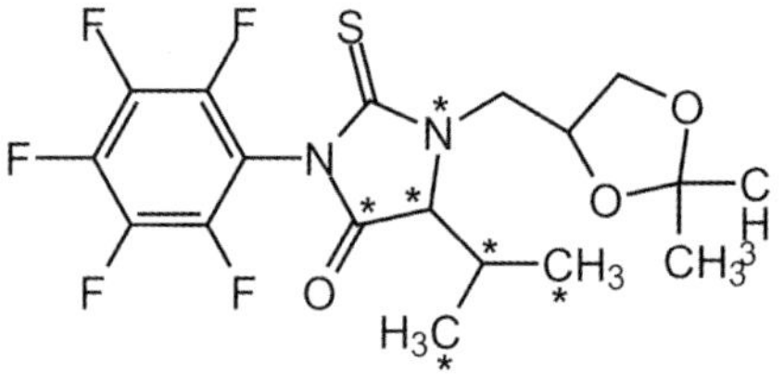

Fig. 2 Ketal derivative of DHPV (* shows the position of the ^{13}C atoms and the ^{15}N atom in the IS ^{13}C$_5$,^{15}N-DHPV).

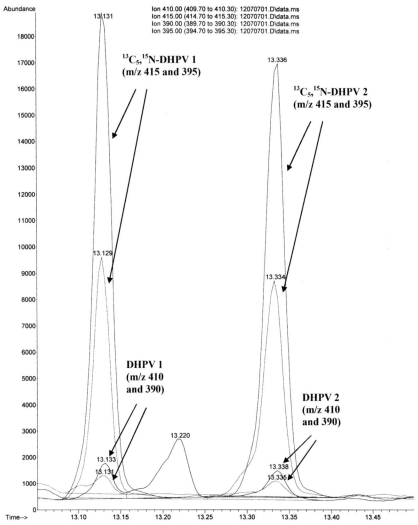

Fig. 3 Ion chromatogram of a calibration standard with 50 pmol DHPV/g globin.

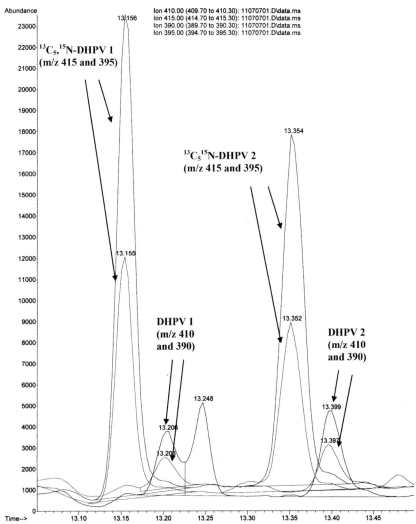

Fig. 4 Ion chromatogram after incubation of a non-smoker haemolysate with 100 µM racemic glycidol.

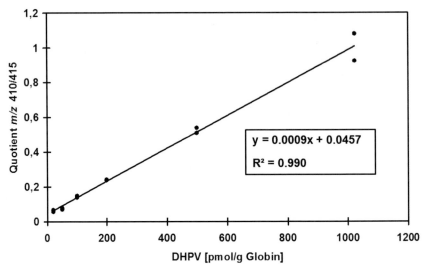

Fig. 5 Example of a calibration graph for the determination of DHPV.

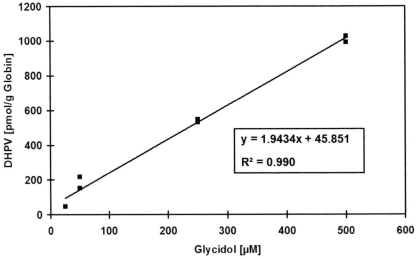

Fig. 6 DHPV adduct formation after incubation of a non-smoker haemolysate with increasing concentrations of racemic glycidol (25–500 µM).

Mercapturic acids
(N-acetyl-S-(2-carbamoylethyl)-L-cysteine, N-acetyl-S-(2-hydroxyethyl)-L-cysteine, N-acetyl-S-(3-hydroxypropyl)-L-cysteine, N-acetyl-S-(2-hydroxypropyl)-L-cysteine, N-acetyl-S-(N-methylcarbamoyl)-L-cysteine) in urine

Matrix:	Urine
Hazardous substances:	Acrylamide, ethylene oxide, acrolein, propylene oxide, N,N-dimethylformamide
Analytical principle:	Liquid chromatography with tandem mass spectrometric detection (LC-MS/MS)
Completed in:	May 2010

Overview of parameters that can be determined with this method and the corresponding hazardous substances:

Substance	CAS	Parameter	CAS
Acrylamide	79-06-1	N-Acetyl-S-(2-carbamoylethyl)-L-cysteine (synonym: N-Acetyl-S-(2-carbonamidethyl)-L-cysteine)	81690-92-8
Ethylene oxide	75-21-8	N-Acetyl-S-(2-hydroxyethyl)-L-cysteine	15060-26-1
Acrolein (2-propenal)	107-02-8	N-Acetyl-S-(3-hydroxypropyl)-L-cysteine	23127-40-4
Propylene oxide (1,2-epoxypropane)	75-56-9	N-Acetyl-S-(2-hydroxypropyl)-L-cysteine	923-43-3
N,N-Dimethylformamide	68-12-2	N-Acetyl-S-(N-methylcarbamoyl)-L-cysteine	103974-29-4

Summary

The present method is used to determine the concentration levels of the mercapturic acids of acrylamide (AAMA), ethylene oxide (HEMA), acrolein (3-HPMA), propylene oxide (2-HPMA) and of dimethylformamide (AMCC) in the urine of potentially exposed workers or individuals of the general population.

The mercapturic acids are extracted from the urine using solid phase extraction, separated by high performance liquid chromatography (HPLC) and detected using tandem mass spectrometry.

Calibration is carried out with calibration standards prepared in pooled urine that are treated in the same manner as the samples to be analysed. As internal standards, isotopically labelled analogues of the analytes (d_4-AAMA, d_4-HEMA, d_3-3-HPMA, $^{13}C_2$-2-HPMA and d_3-AMCC) are added to the urine samples.

N-Acetyl-S-(2-carbamoylethyl)-L-cysteine (AAMA)

Within day precision:	Standard deviation (rel.)	s_w = 5.6% or 2.2%
	Prognostic range	u = 14.4% or 5.7%
	at a spiked concentration of 40 or 400 µg AAMA per litre urine and where n = 6 determinations	
Day to day precision:	Standard deviation (rel.)	s_w = 11.1% or 4.7%
	Prognostic range	u = 27.2% or 11.5%
	at a spiked concentration of 40 or 400 µg AAMA per litre urine and where n = 7 determinations	
Accuracy:	Recovery rate (rel.)	r = 110% or 100%
	at a spiked concentration of 40 or 400 µg AAMA per litre urine and where n = 6 determinations	
Detection limit:	5.0 µg AAMA per litre urine	
Quantitation limit:	15 µg AAMA per litre urine	

N-Acetyl-S-(2-hydroxyethyl)-L-cysteine (HEMA)

Within day precision:	Standard deviation (rel.)	s_w = 5.8% or 4.7%
	Prognostic range	u = 14.9% or 12.1%
	at a spiked concentration of 40 or 400 µg HEMA per litre urine and where n = 6 determinations	
Day to day precision:	Standard deviation (rel.)	s_w = 7.3% or 5.8%
	Prognostic range	u = 17.9% or 14.2%
	at a spiked concentration of 40 or 400 µg HEMA per litre urine and where n = 7 determinations	
Accuracy:	Recovery rate (rel.)	r = 97% or 103%
	at a spiked concentration of 40 or 400 µg HEMA per litre urine and where n = 6 determinations	
Detection limit:	0.5 µg HEMA per litre urine	
Quantitation limit:	1.5 µg HEMA per litre urine	

N-Acetyl-S-(3-hydroxypropyl)-L-cysteine (3-HPMA)

Within day precision:	Standard deviation (rel.)	s_w = 3.1% or 2.6%
	Prognostic range	u = 8.0% or 6.7%
	at a spiked concentration of 40 or 400 µg 3-HPMA per litre urine and where n = 6 determinations	
Day to day precision:	Standard deviation (rel.)	s_w = 20.3% or 8.2%
	Prognostic range	u = 49.7% or 20.1%
	at a spiked concentration of 40 or 400 µg 3-HPMA per litre urine and where n = 7 determinations	
Accuracy:	Recovery rate (rel.)	r = 122% or 131%
	at a spiked concentration of 40 or 400 µg 3-HPMA per litre urine and where n = 6 determinations	
Detection limit:	1.0 µg 3-HPMA per litre urine	
Quantitation limit:	3.0 µg 3-HPMA per litre urine	

N-Acetyl-S-(2-hydroxypropyl)-L-cysteine (2-HPMA)

Within day precision:	Standard deviation (rel.)	s_w = 5.5% or 6.0%
	Prognostic range	u = 14.1% or 15.4%
	at a spiked concentration of 40 or 400 µg 2-HPMA per litre urine and where n = 6 determinations	
Day to day precision:	Standard deviation (rel.)	s_w = 8.6% or 5.2%
	Prognostic range	u = 21.0% or 12.7%
	at a spiked concentration of 40 or 400 µg 2-HPMA per litre urine and where n = 7 determinations	
Accuracy:	Recovery rate (rel.)	r = 109% or 104%
	at a spiked concentration of 40 or 400 µg 2-HPMA per litre urine and where n = 6 determinations	
Detection limit:	1.0 µg 2-HPMA per litre urine	
Quantitation limit:	3.0 µg 2-HPMA per litre urine	

N-Acetyl-S-(N-methylcarbamoyl)-L-cysteine (AMCC)

Within day precision:	Standard deviation (rel.)	s_w = 5.3% or 3.5%
	Prognostic range	u = 13.6% or 9.0%
	at a spiked concentration of 40 or 400 µg AMCC per litre urine and where n = 6 determinations	
Day to day precision:	Standard deviation (rel.)	s_w = 12.8% or 9.4%
	Prognostic range	u = 31.3% or 23.0%
	at a spiked concentration of 40 or 400 µg AMCC per litre urine and where n = 7 determinations	
Accuracy:	Recovery rate (rel.)	r = 117%
	at a spiked concentration of 40 or 400 µg AMCC per litre urine and where n = 6 determinations	
Detection limit:	5.0 µg AMCC per litre urine	
Quantitation limit:	15 µg AMCC per litre urine	

Mercapturic acids as elimination products of hazardous substances

Acrylamide Acrylamide has been classified by the German Research Foundation (DFG, *Deutsche Forschungsgemeinschaft*) in Category 2 of carcinogenic substances [1]. Industrially, acrylamide is mainly employed for the production of polyacrylamides which are used as dispersants and flocculants for the treatment of drinking water. In addition, high molecular polyacrylamides can be modified chemically by introducing non-ionic, anionic or cationic groups for various purposes. These are used in the paper industry as ion exchangers, thickeners or processing agents. Furthermore, acrylamide is used in the production of paints, as copolymer for various plastics (including the manufacture of fire-resistant glass) and as sealant in the construction industry. In research, acrylamide is used to pour polyacrylamide gels for electrophoresis [2].

In April 2002, a Swedish working group demonstrated that acrylamide is formed in various starch-containing foods with a low water content when heated to high temperatures [3]. From that it can be concluded that exposure to acrylamide – even when avoiding foods such as chips and potato crisps – can scarcely be avoided. The general population is thus exposed to acrylamide on a broad basis via the consumption of foods such as bakery products, bread and coffee [1].

The toxicity of acrylamide is comprehensively described in the 2007 MAK documentation [1]. Acrylamide is rapidly absorbed orally, dermally and via inhalation and, on account of its high solubility in water, distributed throughout the whole body and immediately metabolised, mainly to the glutathione conjugate. In rat studies approx. 25% of a dermally applied acrylamide dose was absorbed within 24 hours, thus confirming its efficient absorption through the skin [4].

During biotransformation, the initially formed glutathione conjugates are further metabolised and eliminated in the urine. Binding of glutathione to acrylamide occurs directly via a Michael addition and results in the formation of N-acetyl-S-(2-carbamoylethyl)-L-cysteine (AAMA) [5]. After oral administration of acrylamide to humans, about 52% of an administered dose is eliminated in the form of AAMA in the urine within 48 hours. After a remarkably long distribution phase in the body of up to 18 hours following administration, AAMA is eliminated in two phases with half-lives of 3.5 hours and more than 10 hours, respectively [6]. In addition, acrylamide is also able to form covalently bound adducts with circulating blood proteins such as haemoglobin (see Figure 1). These adducts are also analytically accessible by biochemical effect monitoring [2].

It is known that the biotransformation of acrylamide after epoxidation of the double bond by cytochrome P 450 monooxygenases may lead to the formation of the epoxide glycidamide. Glycidamide can also form a haemoglobin adduct or be eliminated as mercapturic acid in the urine after conjugation with glutathione [1].

Up to now, however, biological threshold and reference values by the DFG only exist for the carbamoylethyl compounds as direct metabolites of acrylamide. For the haemoglobin adduct N-2-carbamoylethyl valine (see Figure 1) a *"Biologischer Leitwert"* (BLW) of 550 pmol/g globin, a *"Biologischer Arbeitsstoff-Referenzwert"* (BAR) of 50 pmol/g globin and an *EKA Correlation* (Exposure Equivalents for Car-

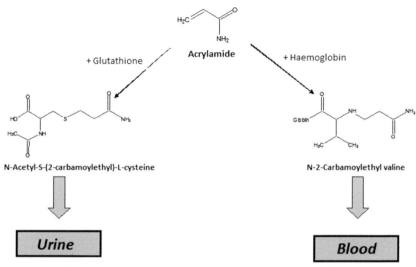

Fig. 1 Simplified metabolic scheme of acrylamide.

cinogenic Substances) were assessed [7]. Since 2011 another BAR value has been set for the corresponding mercapturic acid AAMA at a level of 100 µg/g creatinine [7].

Furthermore, the Committee on Hazardous Substances (*Ausschuss für Gefahrstoffe*, AGS) has derived an exposure-risk relationship (*Expositions-Risiko-Beziehung*, ERB) for acrylamide levels in the air at the workplace [8].

Ethylene oxide At room temperature, ethylene oxide is a colourless, highly inflammable gas with a faint sweet ether-like odour and has no reliable sensory warning properties at a mean odour threshold of 700 mL/m^3 [9, 10].

The substance is an intermediate in the production of several chemicals. It is mainly used in the production of ethylene glycol (as antifreeze or cooling agent). In addition, it is used to make polyesters (e.g. PET – polyethylene terephthalate).

A further application field for ethylene oxide is the sterilisation of medical equipment and materials. Here, PVC packaged medical devices are disinfected with gaseous ethylene oxide in an enclosed space. Ethylene oxide penetrates the packaging and sterilises the materials due to its biocidal properties. Other application fields for ethylene oxide are the production of surfactants, special detergents and emulsifiers, where starting materials such as phenols or fatty acids are converted with alkylating substances. Furthermore, tobacco smoke contains ethylene oxide, meaning that the internal exposure is higher in smokers than in non-smokers. However, a low physiological background level is even found in non-smokers [11, 12].

The toxicity of ethylene oxide is comprehensively described in the MAK value documentations of 1984 and 2002 [9, 10]. It is classified in Category 2 of carcino-

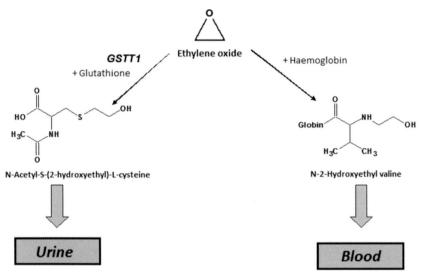

Fig. 2 Simplified metabolic scheme of ethylene oxide.

genic substances and is additionally designated with an "H" due to its possible dermal absorption at the workplace [7]. As electrophilic alkylating agent ethylene oxide reacts in the body with nucleophilic groups such as carboxyl, amino, phenolic hydroxyl and sulfhydryl groups. The general biocidal properties of ethylene oxide are due to its high reactivity [9, 10].

Ethylene oxide can be detoxified in the body by conjugation with glutathione, which results in elimination of the corresponding mercapturic acid N-acetyl-S-(2-hydroxyethyl)-L-cysteine (HEMA) [13]. Glutathione conjugation is catalysed by the enzyme glutathione-S-transferase GSTT1 [14] (see Figure 2).

Due to the high reactivity of ethylene oxide it is able – like other carcinogenic substances – to react with DNA and proteins, such as haemoglobin for example (see Figure 2). The corresponding N-terminal valine adduct can be quantified by biological effect monitoring. For ethylene oxide and ethylene an *EKA Correlation* has been established which describes a linear relationship between external exposure to ethylene or ethylene oxide and the concentration level of the haemoglobin adduct hydroxyethyl valine in blood [7]. An exposure-risk relationship (*Expositions-Risiko-Beziehung*, ERB), has additionally been derived for ethylene oxide by the Committee on Hazardous Substances (*Ausschuss für Gefahrstoffe*, AGS). Hereby, an air concentration of 0.1 ppm was set as acceptable risk and an air concentration of 1 ppm as tolerable risk [8]. According to the *EKA Correlation* the latter corresponds to an internal exposure of 90 µg of the Hb adduct N-hydroxyethyl-valine per litre blood.

Acrolein (2-propenal) Acrolein is an α,β-unsaturated aldehyde showing high reactivity. Industrially, it is mainly used in the production of acrylic acid and acrylates as well as glycerol. In addition, acrolein is found in considerable quantities in the exhaust fumes of combustion engines as well as in cigarette smoke. It is also a natural component of various foods as fruit, vegetables or red wine. Physiologically, acrolein is also produced during the degradation of fatty acids. Accordingly, humans are exposed to acrolein – both exogenously and endogenously – from innumerable sources [15]. A detailed description of the toxicity of acrolein is given in the 1997 MAK documentation [16]. Acrolein is classified in Category 3B of carcinogenic substances [7].

During its metabolism, acrolein is conjugated with glutathione. By subsequent reduction (by aldehyde reductases), 3-hydroxypropyl mercapturic acid (3-HPMA) is formed, which is eliminated in the urine (see Figure 3).

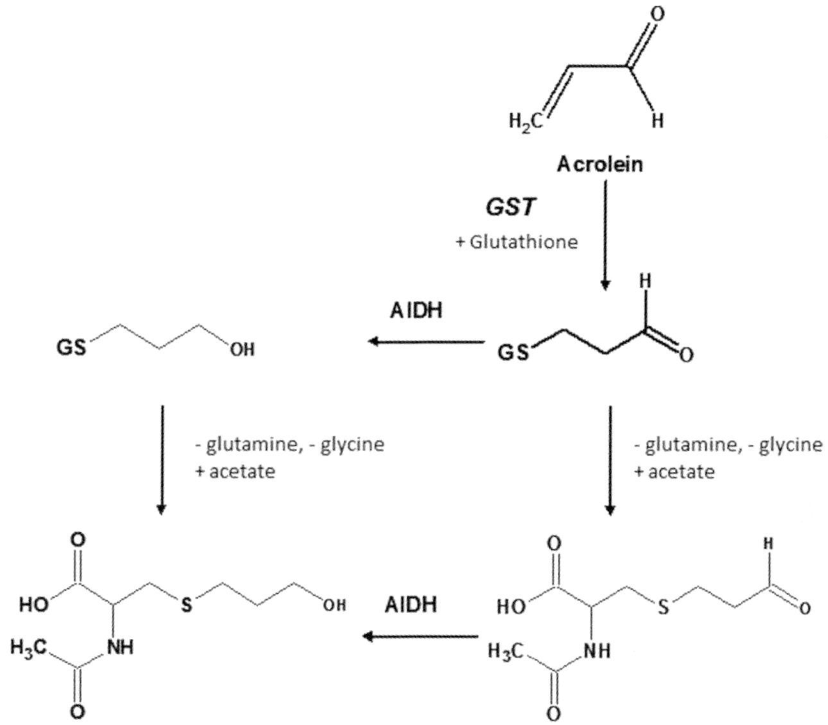

N-Acetyl-S-(3-hydroxypropyl)-L-cysteine

Fig. 3 Simplified metabolism of acrolein involving the formation of 3-HPMA.

Propylene oxide (1,2-epoxypropane) Propylene oxide is produced industrially using the chlorohydrin process on large-scale in quantities of about 3.6 million tonnes per year (1998). It serves as starting material in the production of propylene glycol, propylene glycol ethers, alkyl cellulose ethers, isopropanol amines and surfactants. At room temperature, propylene oxide is a colourless liquid with a faint sweet odour and a very high vapour pressure. In air, the odour threshold is 100–350 ppm; this parameter should therefore not be used as an indicator for propylene oxide in ambient air [17]. Propylene oxide is percutaneously absorbed and has therefore been designated with an "H" by the DFG [7]. The substance has been classified in Category 2 of carcinogenic substances by the DFG [7].

In the course of its metabolism – presumably also by glutathione transferases – propylene oxide is conjugated with glutathione and excreted in the urine in the form of N-acetyl-(2-hydroxypropyl)-L-cysteine (2-HPMA) (see Figure 4). In 2011 a "Biologischer Arbeitsstoff-Referenzwert" (BAR) was set for propylene oxide at the level of 25 µg 2-HPMA per g creatinine [7]. On account of its high reactivity, propylene oxide can also react with haemoglobin. However, in direct comparison with ethylene oxide, its reactivity is lower by a factor of approx. 5 [18].

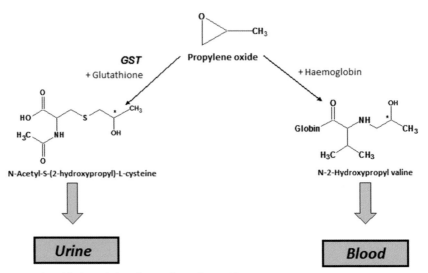

Fig. 4 Simplified metabolic scheme of propylene oxide.

N,N-Dimethylformamide N,N-Dimethylformamide (DMF) is a colourless polar organic solvent with a boiling point of 153°C. On account of its practically unlimited miscibility with water and other organic compounds, it is a popular solvent in the pharmaceutical and chemical industry.

In the human body, DMF is metabolised to N-hydroxymethyl-N-methylformamide (HMMF), N-methylformamide (NMF), N-acetyl-(N-methylcarbamoyl)-cysteine (AMCC) and a series of other metabolites, which are eliminated in the urine [19]. A detailed summary of its toxicity can be found in the MAK documentations on DMF [20, 21]. The current MAK value of DMF in the air is 5 ppm (15 mg/m³). DMF is additionally designated with an "H" due to its possible dermal absorption at the workplace. Furthermore, DMF has been classified in pregnancy risk group B [7].

Its BAT value is set at 35 mg per litre urine for the sum of HMMF and NMF, measured at the end of a shift [7]. For the mercapturic acid of DMF, AMCC, the Biological Exposure Index (BEI) in the USA is set at 40 mg/L (measured in pre-shift urine samples at the end of a working week).

General Preliminary Remarks Up to now, haemoglobin adduct determination for the listed industrial chemicals has mainly been used for biomonitoring studies reflecting the cumulative long-term dose, whereas the mercapturic acids are short-term biomarkers reflecting the exposure during the last hours or days before specimen collection. The presented method is therefore suitable for short-term monitoring of the effectiveness of occupational hygiene measures.

The method was applied in initial studies using urine samples from a small group of smokers and non-smokers of the general population and has meanwhile been internationally published [22]. Differentiation between smokers and non-smokers has been carried out using cotinine analyses in the urine. As Table 1 shows, the results obtained with the present method are comparable with those in other international studies on the elimination of mercapturic acids in the general population.

Table 1 Literature data on the elimination of mercapturic acids in the urine of the general population (NS = non-smokers, S = smokers, occ. S = occasional smokers).

	n	Median (range) [µg/L]					Ref.
		AAMA	HEMA	3-HPMA	2-HPMA	AMCC	
NS	14	52.6 (12.7–70.8)	2.0 (0.7–4.7)	155.0 (37.0–730.4)	7.1 (<1–49.3)	113.6 (38.9–498.3)	[22]
S	14	242.7 (30.3–446.9)	5.3 (0.8–7.5)	1681 (132.4–5345)	41.7 (4.4–251.8)	821.7 (122–1453)	
NS	16	29 (<5–83)	–	–	–	–	[5]
S	13	127 (17–338)	–	–	–	–	
NS	13	26 (14–102)	–	–	–	–	[23]
occ. S	12	56 (16–630)	–	–	–	–	
S	13	283 (61–706)	–	–	–	–	

Table 1 (Continued)

	n	Median (range) [μg/L]					Ref.
		AAMA	HEMA	3-HPMA	2-HPMA	AMCC	
NS	242	–	1.1[a] 6.9[b]	–	–	–	[24]
S	170	–	2.6[a] 15.4[b]	–	–	–	
NS	41	–	–	812[a]	–	–	[25]
S	27	–	–	2809[a]	–	–	
NS	32	–	–	–	–	40.2 (<30–78.2)	[26]
S	10	–	–	–	–	38.7 (32.6–62.7)	

[a] mean value; [b] 95th percentile

Author: *Th. Schettgen*
Examiners: *G. Scherer, K. Sterz*

Mercapturic acids
(N-acetyl-S-(2-carbamoylethyl)-L-cysteine, N-acetyl-S-(2-hydroxyethyl)-L-cysteine, N-acetyl-S-(3-hydroxypropyl)-L-cysteine, N-acetyl-S-(2-hydroxypropyl)-L-cysteine, N-acetyl-S-(N-methylcarbamoyl)-L-cysteine) in urine

Matrix:	Urine
Hazardous substances:	Acrylamide, ethylene oxide, acrolein, propylene oxide, N,N-dimethylformamide
Analytical principle:	Liquid chromatography with tandem mass spectrometric detection (LC-MS/MS)
Completed in:	May 2010

Contents

1	General principles
2	Equipment, chemicals and solutions
2.1	Equipment
2.2	Chemicals
2.3	Solutions
2.4	Internal standards
2.5	Calibration standards
3	Specimen collection and sample preparation
3.1	Specimen collection
3.2	Sample preparation
4	Operational parameters
4.1	Operational parameters for high-performance liquid chromatography
4.2	Operational conditions for mass spectrometry

5	Analytical determination
6	Calibration
7	Calculation of the analytical result
8	Standardisation and quality control
9	Evaluation of the method
9.1	Precision
9.2	Accuracy
9.3	Absolute recovery
9.4	Matrix effects
9.5	Stability of the analytes
9.6	Detection limits and quantitation limits
9.7	Sources of error
10	Discussion of the method
11	References
12	Appendix

1 General principles

The mercapturic acids are extracted from urine using solid phase extraction, separated by HPLC and subsequently detected using tandem mass spectrometry.

Calibration is carried out with calibration standards prepared in pooled urine that are treated in the same manner as the samples for analysis. As internal standards isotopically labelled analogues of the analytes (d_4-AAMA, d_4-HEMA, d_3-3-HPMA, $^{13}C_2$-2-HPMA and d_3-AMCC) are added to the urine samples.

2 Equipment, chemicals and solutions

2.1 Equipment

- HPLC-MS/MS system consisting of a quaternary gradient pump (e.g. Agilent 1100 series), a degassing device, a column thermostat, an injection valve and a sample loop for 100 µL, an autosampler, a 6-port valve, a tandem mass spectrometric detector (e.g. Applied Biosystems API 3000) and a PC system for data evaluation
- HPLC column: Luna C-8 (2), length: 150 mm; inner diameter: 4.6 mm; particle diameter: 3 µm (Phenomenex, No. 00F-4248-E0)
- Precolumn: C-8, length 3 mm; inner diameter: 4 mm; particle diameter 3 µm (e.g. Phenomenex, No. AJ0-4290)
- Precolumn filter: 3 mm x 0.5 µm for HPLC columns with 4.6 mm diameter (e.g. Supelco, No. 57677)

- Cartridges for solid phase extraction: Isolute ENV+, 100 mg, 3 mL (e.g. Biotage, Uppsala, Sweden)
- pH Meter (e.g. Mettler-Toledo)
- Magnetic stirrer
- Workstation for solid phase extraction with vacuum pump (e.g. Baker)
- Laboratory shaker (e.g. IKA Vibrax VXR)
- Evaporation unit using nitrogen
- 10 mL and 25 mL Screw cap glass tubes
- 1.8 mL Screw cap vials with PTFE septa (e.g. Macherey-Nagel)
- Microlitre pipettes, with adjustable volumes between 10 and 100 µL as well as between 100 and 1000 µL (e.g. Eppendorf)
- Various volumetric flasks
- Various glass beakers
- Polypropylene containers for urine collection (e.g. Kautex)
- Centrifuge (e.g. Hettich, Modell Rotina 420R)

2.2 Chemicals

Unless otherwise specified, all listed chemicals must be at least of p.a. quality.

- N-Acetyl-S-(2-carbamoylethyl)-L-cysteine (AAMA, e.g. Dr. Ehrenstorfer, No. X-6188)
- d_4-N-Acetyl-S-(2-carbamoylethyl)-L-cysteine (d_4-AAMA, e.g. Dr. Ehrenstorfer, No. D-6589)
- N-Acetyl-S-(2-hydroxyethyl)-L-cysteine (HEMA, e.g. Toronto Research Chemicals, No. A178910)
- d_4-N-Acetyl-S-(2-hydroxyethyl)-L-cysteine (d_4-HEMA, e.g. Toronto Research Chemicals, No. A178913)
- N-Acetyl-S-(3-hydroxypropyl)-L-cysteine (3-HPMA, e.g. Toronto Research Chemicals, No. A179070)
- d_3-N-Acetyl-S-(3-hydroxypropyl)-L-cysteine, dicyclohexylammonium salt (d_3-3-HPMA, e.g. Toronto Research Chemicals, No. A179077)
- N-Acetyl-S-(2-hydroxypropyl)-L-cysteine dihydrate (2-HPMA, e.g. Toronto Research Chemicals, No. A179060)
- $^{13}C_2$-N-Acetyl-S-(2-hydroxypropyl)-L-cysteine ($^{13}C_2$-2-HPMA, custom synthesis, provided by the Institute of Occupational, Social and Environmental Medicine of the University of Erlangen-Nuremberg)
- N-Acetyl-S-(N-methylcarbamoyl)-cysteine (AMCC, e.g. Fluka, No. 90914)
- d_3-N-Acetyl-S-(N-methylcarbamoyl)-cysteine (d_3-AMCC, e.g. Toronto Research Chemicals, No. A186627)
- Formic acid (100%) (e.g. Merck, No. 1.00264)
- Acetic acid (100%) (e.g. Merck, No. 1.00063)
- Methanol (e.g. Merck, No. 1.06007)

- Ammonium formiate (e.g. Fluka, No. 55674)
- Acetonitrile, HPLC grade (e.g. Baker, No. 98.212.500)
- Bidistilled water (e.g. from Millipore™ technology)
- Nitrogen 5.0 (e.g. Linde)

2.3 Solutions

- Ammonium formiate buffer (50 mM), pH = 2.5
  788 mg ammonium formiate are weighed exactly into a 250 mL glass beaker and are dissolved in approx. 200 mL bidist. water. Subsequently, the solution is adjusted to a pH value of 2.5 with formic acid under stirring and continuous pH value monitoring using a pH meter. The solution is transferred to a 250 mL volumetric flask, which is made up to the mark with bidist. water.
- 0.1% Formic acid (pH = 2.5)
  500 µL formic acid are pipetted into a 500 mL volumetric flask and approx. 450 mL bidist. water are added. The pH value of the solution is adjusted to pH = 2.5 by dropwise addition of formic acid under stirring on a magnetic stirrer and continuous monitoring using a pH meter. The flask is then made up to the mark with bidist. water.
- Wash solution for solid phase extraction (0.1% formic acid (pH = 2.5)/methanol (9:1, v:v))
  10 mL methanol are pipetted into a 100 mL volumetric flask, which is then made up to the mark with 0.1% formic acid (pH = 2.5).
- Elution solution for solid phase extraction (1% formic acid in methanol)
  1 mL formic acid is pipetted into a 100 mL volumetric flask and the flask made up to the mark with methanol.
- 1% Formic acid
  1 mL formic acid is pipetted into a 100 mL volumetric flask and the flask is made up to the mark with bidist. water.

Using cool storage (4°C) conditions, the solutions can be kept for at least one month.

- Eluent A (0.1% formic acid, pH = 2.5)
  Preparation see above.
- Eluent B (60% eluent A/40% acetonitrile)
  300 mL eluent A are placed in a 500 mL volumetric flask. The volumetric flask is then made up to the mark with acetonitrile.

2.4 Internal standards

- Stock solution d_4-AAMA (1 g/L)
  10 mg d_4-N-acetyl-(2-carbamoylethyl)-L-cysteine (d_4-AAMA) are weighed exactly into a 10 mL volumetric flask, which is then made up to the mark with bidist. water.
- Stock solution d_4-HEMA (1 g/L)
  1 mg d_4-N-acetyl-(2-hydroxyethyl)-L-cysteine (d_4-HEMA) is dissolved in 1 mL bidist. water in a 1.8 mL vial.
- Stock solution d_3-3-HPMA (550 mg/L)
  1 mg d_3-N-acetyl-(3-hydroxypropyl)-L-cysteine dicyclohexylammonium salt (d_3-3-HPMA) is dissolved in 1 ml bidist. water in a 1.8 mL vial. Taking in consideration the molar ratio to dicyclohexylammonium salt, the concentration of the solution is 550 mg/L.
- Stock solution d_3-AMCC (1 g/L)
  10 mg d_3-N-acetyl-S-(N-methylcarbamoyl)-cysteine (d_3-AMCC) are weighed exactly into a 10 mL volumetric flask, which is then made up to the mark with bidist. water.
- Stock solution $^{13}C_2$-2-HPMA (500 mg/L)
  5 mg $^{13}C_2$-N-Acetyl-S-(2-hydroxypropyl)-L-cysteine ($^{13}C_2$-2-HPMA) are weighed exactly into a 10 mL volumetric flask, which is then made up to the mark with bidist. water.
- Working solution of the internal standards:
  100 µL each of the stock solutions of the internal standards (d_4-AAMA, d_4-HEMA, d_3-3-HPMA, d_3-AMCC and $^{13}C_2$-2-HPMA) are pipetted into a 10 mL volumetric flask which is then made up to the mark with 0.1% formic acid. The concentrations of the working solution of the internal standards are 10 mg/L each for d_4-AAMA, d_4-HEMA and d_3-AMCC, 5.5 mg/L for d_3-3-HPMA, and 5 mg/L for $^{13}C_2$-2-HPMA.

The solutions are stored at –20°C in brown screw cap vials and can thus be kept without noteworthy losses for at least 6 months.

2.5 Calibration standards

- Stock solution AAMA (1 g/L)
  10 mg N-acetyl-(2-carbamoylethyl)-L-cysteine (AAMA) are weighed exactly into a 10 mL volumetric flask, which is then made up to the mark with bidist. water.
- Stock solution HEMA (1 g/L)
  10 mg N-acetyl-(2-hydroxyethyl)-L-cysteine (HEMA) are weighed exactly into a 10 mL volumetric flask. The flask is then made up to the mark with bidist. water.

- Stock solution 3-HPMA (1.1 g/L)
  10 mg N-acetyl-(3-hydroxypropyl)-L-cysteine dicyclohexylammonium salt (3-HPMA) are weighed exactly into a 5 mL volumetric flask. The volumetric flask is then made up to the mark with bidist. water. Taking the molar ratio to dicyclohexylammonium salt into account the concentration of the solution is 1.1 g 3-HPMA per litre.
- Stock solution 2-HPMA (0.85 g/L):
  10 mg N-acetyl-(2-hydroxypropyl)-L-cysteine dihydrate (2-HPMA) are weighed exactly into a 10 mL volumetric flask. The volumetric flask is then made up to the mark with bidist. water. Taking the molar ratio into account, the concentration of the solution is 0.85 g 2-HPMA per litre.
- Stock solution AMCC (1 g/L):
  10 mg N-acetyl-S-(N-methylcarbamoyl)-cysteine (AMCC) are weighed exactly into a 10 mL volumetric flask, which is then made up to the mark with bidist. water.
- Working solution 1 (100 mg/L):
  1 mL each of the stock solutions of the mercapturic acids (AAMA, HEMA, 3-HPMA, 2-HPMA and AMCC) are pipetted into a 10 mL volumetric flask which is then made up to the mark with 0.1% formic acid. The concentration of working solution 1 is approx. 100 mg/L per analyte.
- Working solution 2 (10 mg/L):
  100 µL each of the stock solutions of the mercapturic acids (AAMA, HEMA, 3-HPMA, 2-HPMA and AMCC) are pipetted into a 10 mL volumetric flask which is then made up to the mark with 0.1% formic acid. The concentration of working solution 2 is approx. 10 mg/L per analyte.

The solutions are stored at –20°C in brown screw cap vials and can thus be kept without noteworthy losses for at least 6 months.

The calibration standard solutions are prepared in acidified filtered pooled urine of individuals from the general population that are occupationally non-exposed to the respective alkylating substances. In order to keep the background level of the mercapturic acids in the pooled urine used as low as possible, urine of non-smokers should be used exclusively. To prepare the pooled urine, spontaneous urine samples from the participants are collected in a suitable container, mixed thoroughly and stored at –20°C until preparation of the standards and the control material. After thawing, the pooled urine is passed through a folded filter in order to separate precipitated proteins. Following this, the urine is acidified with glacial acetic acid to 1% acetic acid (v:v) and can then be used to prepare the standards.

Calibration standards in a concentration range between 10 and 4000 µg/L are prepared by dilution of the working solutions with pooled urine according to the pipetting scheme given in Table 2. Unspiked pooled urine is included as blank value. If the method is to be applied for determination of the background levels of the mercapturic acids, it is advisable to carry out calibration for the analytes HEMA and 2-HPMA within a smaller concentration range (e.g. HEMA: 1 to 50 µg/L; 2-HPMA: 2.5 to 250 µg/L), as the urinary levels to be expected from occupationally

Table 2 Pipetting scheme for the preparation of calibration standards in pooled urine.

Calibration standard	working solution 1 [µL]	working solution 2 [µL]	Final volume [mL]	Analyte level [µg/L]
K1	–	25	25	10
K2	–	125	25	50
K3	–	250	25	100
K4	–	500	25	200
K5	125	–	25	500
K6	500	–	25	2000
K7	1000	–	25	4000

non-exposed individuals are markedly lower than that of the other analytes (see also Table 1).

The calibration standards are aliquoted into quantities of 2.1 mL each and stored at –20°C. Thus, the standard solutions can be kept without noteworthy losses for at least 6 months.

3 Specimen collection and sample preparation

3.1 Specimen collection

The urine samples are collected in sealable plastic containers and kept at –20°C. Thus, the samples can be kept without noteworthy losses for at least 6 months.

3.2 Sample preparation

Prior to analysis, the urine samples are thawed at room temperature and mixed thoroughly. 2 mL of the sample are pipetted into a 25 mL screw cap vial and 2 mL of the ammonium formiate buffer (pH = 2.5) as well as 40 µL formic acid are added. The pH value of the urine sample then ought to be pH = 2.5 and can be checked if necessary using pH paper. Subsequently, 40 µL of the working solution of the internal standards are added. The samples are mixed thoroughly on a laboratory shaker and are then centrifuged at 1400 g for 10 min.

For solid phase extraction, the cartridges are first preconditioned. For this purpose, the cartridges are consecutively rinsed with 4 mL methanol, 2 mL bidist. water and 2 mL 0.1% formic acid (pH = 2.5). Subsequently, the complete supernatant of the centrifuged sample solution is transferred onto the cartridge and is allowed to drop slowly through the column. Then, the cartridge is rinsed twice consecutively with 500 µL 0.1% formic acid (pH = 2.5) as well as with 400 µL of the wash solution. These working steps should be done without using a vacuum pump if possible!

Thereafter, the cartridge is sucked dry using a vacuum pump for at least 45 min and is then centrifuged (2500 g, 5 min) to remove any residual water. The solid phase station is also cleared of water drops in the meantime, whereby the stopcocks should also be checked (by tapping!). After centrifugation the cartridge is again fitted onto the solid phase station and sucked dry again for at least 30 min.

Elution of the analytes is subsequently carried out three times with 600 µL each of the elution solution into a 10 mL glass vial. Here too, elution is to be carried out without using the vacuum pump as far as possible. The eluate is transferred to a 1.8 mL screw cap vial and the solution is carefully evaporated under a stream of nitrogen. This process can be accelerated by slight heating (lukewarm only!).

The residue is subsequently dissolved in 500 µL 0.1% formic acid (pH = 2.5) (rinse glass walls). Then, the screw cap vials are sealed and briefly shaken.

4 Operational parameters

The analytical determination is carried out using an equipment setup consisting of an HPLC gradient pump Agilent 1100 series, a 6-port valve and an Applied Biosystems API-3000 LC-MS/MS system.

4.1 Operational parameters for high-performance liquid chromatography

Analytical column:	Material:	Silica gel
	Length:	150 mm
	Inner diameter:	4.6 mm
	Column filling:	Phenomenex Luna C8 (2), 3 µm
Separation principle:	Reversed phase	
Column temperature:	35°C	
Detection:	Tandem mass spectrometric detector	
Mobile phase:	Eluent A:	0.1% formic acid pH = 2.5
	Eluent B:	60% eluent A/40% acetonitrile
Gradient:	See Table 3	
Flow rate:	0.3 mL/min	
Injection volume:	100 µL	

All other parameters must be optimised in accordance with the manufacturer's instructions.

Table 3 Gradient pump program.

Time (min)	Eluent A vol.%	Eluent B vol.%
0	75	25
2	75	25
8	50	50
12	50	50
14	0	100
18	0	100
22	75	25
26	75	25

4.2 Operational conditions for mass spectrometry

Ion source settings:

Ionisation mode:	ESI negative
Source temperature:	430°C
Ion spray voltage:	–4500 V
Nebulizing Gas (NEB):	Nitrogen, 10 psi
Curtain gas (CUR):	Nitrogen, 6 psi
Collision gas (CAD):	Nitrogen, 6 psi

Quadrupol settings:

Q 1 Resolution:	Unit
Q 3 Resolution:	Unit
Scan type:	MRM (Multi-Reaction-Mode)
Dwell time:	300 msec
Parameter specific settings:	See Table 4

Table 4 Parameter specific setting at the MS/MS (DP – declustering potential, FP – focussing potential, EP – entrance potential, CE – collision energy, CXP – collision exit potential).

	Q 1 [m/z]	Q 3 [m/z]	DP [V]	FP [V]	EP[V]	CE[V]	CXP [V]
AAMA	232.862	103.950	–81	–180	–10	–24	–5
d$_4$-AAMA	236.933	107.950	–81	–240	–10	–20	–5
HEMA	205.992	77.100	–16	–190	–10	–20	–4
d$_4$-HEMA	209.992	81.100	–16	–190	–10	–20	–4
3-HPMA	220.030	91.000	–21	–240	–10	–20	–6
2-HPMA	220.030	91.000	–21	–240	–10	–20	–6
d$_3$-3-HPMA	222.994	90.900	–16	–150	–10	–20	0
^{13}C$_2$-2-HPMA	221.985	92.950	–56	–50	–10	–18	–5
AMCC	218.952	162.000	–21	–280	–10	–12	0
d$_3$-AMCC	222.002	164.900	–1	–170	–10	–20	–6

The analytical conditions are given for the equipment configuration used and must be optimised according to instructions where other manufacturers' equipment is used.

To avoid contamination of the ion source and to reduce resultant exposure of the system, the flow, with the help of a 6-port valve, was conducted into the mass spectrometer only in the time window relevant for detection of the mercapturic acids (7.5–13 min). The 6-port valve was controlled by the autosampler.

5 Analytical determination

From the urine samples prepared according to Section 3, 100 µL are injected into the HPLC unit respectively.

The temporal profiles of the ion transitions listed in Table 5 are recorded in the MRM mode of the tandem mass spectrometer (ESI negative mode).

The retention times given in Table 5 are intended to be a rough guide only. Users of the method must ensure proper separation performance of the analytical column used and the resulting retention behaviour of the analytes.

Figures 5–7 (see Appendix) show the chromatogram of a processed urine sample of a non-smoker. The product ion spectra of the individual analytes are shown in the Appendix in Figures 9a–e.

Table 5 Retention times and analysed ion transitions.

Analyte	Retention time [min]	Ion transitions (m/z)	
		Q 1	Q 3
AAMA	8.32	233	104
d_4-AAMA	8.31	237	108
HEMA	8.37	206	77
d_4-HEMA	8.36	210	81
3-HPMA	10.18	220	91
d_3-3-HPMA	10.21	223	91
AMCC	10.34	219	162
d_3-AMCC	10.32	222	165
2-HPMA	11.09	220	91
$^{13}C_2$-2-HPMA	11.09	222	93

6 Calibration

The calibration standards are prepared and processed according to Section 2.5 and Section 3 and are analysed as described in Section 4. For this purpose, 100 µL of the prepared calibration standard solution are injected into the LC-MS/MS system.

A reagent blank value is also included in every analytical series and processed as described in Section 3, whereby bidist. water is used in place of the urine sample.

Calibration graphs are obtained by plotting the quotients of the peak areas of the analytes and the corresponding isotopically labelled internal standard against the nominal concentrations of the calibration standards. Any blank values which may still be present are accounted for by subtraction.

When using a 2 mL sample, linear measurement ranges of 10–2000 µg/L for AAMA and 10–4000 µg/L for HEMA, 3-HPMA and AMCC each were determined. A flattening of the calibration graph at concentration levels above 500 µg/L urine was observed for 2-HPMA only. This is due to mass spectrometric interferences of the analyte with the internal standard as the IS for 2-HPMA is labelled only twice (see also Section 9.7). Exemplary, Figure 8 shows the calibration graphs for the mercapturic acids in pooled urine.

7 Calculation of the analytical result

The analyte concentration in the urine samples is calculated on the basis of the calibration graphs (see Section 6). The peak area of each analyte is divided by the peak area of the respective internal standard. From the quotient thus obtained, the corresponding concentration of the analytes in µg per litre urine is obtained (using the relevant calibration graph). If the determined analyte level of the sample is outside the calibration range, the urine sample is diluted with bidist. water at a ratio of 1:10 and prepared again.

8 Standardisation and quality control

Quality control of the analytical results is carried out as stipulated in the guidelines of the *Bundesärztekammer* (German Medical Association) [27, 28].

For quality control, a quality control sample with a constant analyte concentration is analysed within each analytical series. As a quality control material is not commercially available, it must be prepared in the laboratory.

For this purpose, a defined amount of the mercapturic acids to be analysed is spiked to pooled urine. Aliquots of this solution are stored at −20°C and included as quality control samples within each analytical series. The nominal value and the tolerance ranges of this quality control material are determined in a pre-analytical period (one analysis of the control material on each of 10 different days) [28].

9 Evaluation of the method

9.1 Precision

To determine the within day and the day to day precision pooled urine was spiked with defined quantities of the mercapturic acids to obtain concentration levels of 40 and 400 µg/L, respectively.

These urine samples were prepared and analysed six times in a row to obtain the within day precision documented in Table 6.

To determine the day to day precision, the same sample material was used as for determining the within day precision. The quality control samples were prepared and analysed on seven different days. The resulting precision data are shown in Table 7.

Table 6 Within day precision for the determination of mercapturic acids in urine (n = 6).

Analyte	Mean value [µg/L]	Standard deviation (rel.) [%]	Prognostic range [%]
AAMA	444	2.2	5.7
	88	5.6	14.4
HEMA	428	4.7	12.1
	54	5.8	14.9
3-HPMA	630	2.6	6.7
	154	3.1	8.0
2-HPMA	432	6.0	15.4
	59	5.5	14.1
AMCC	529	3.5	9.0
	107	5.3	13.6

Table 7 Day to day precision for the determination of mercapturic acids in urine (n = 7).

Substance	Mean value [µg/L]	Standard deviation (rel.) [%]	Prognostic range [%]
AAMA	435	4.7	11.5
	71	11.1	27.2
HEMA	391	5.8	14.2
	41	7.3	17.9
3-HPMA	625	8.2	20.1
	166	20.3	49.7
2-HPMA	376	5.2	12.7
	53	8.6	21.0
AMCC	514	9.4	23.0
	118	12.8	31.3

9.2 Accuracy

To determine the influence of matrix effects on the accuracy of the method, a comparison between the calibration graphs in water and in pooled urine was carried out. The obtained slopes of the calibration graphs are shown in Table 8.

Table 8 Slopes of the calibration graphs in water or pooled urine.

Substance	Slope in water	Slope in pooled urine	Deviation [%]
AAMA	0.00237	0.00288	+21
HEMA	0.00509	0.00524	+3.0
3-HPMA	0.00491	0.00527	+7.5
2-HPMA	0.00170	0.00158	−7.0
AMCC	0.00202	0.00183	−9.7

To determine the accuracy of the method, recovery experiments were carried out. For this purpose, pooled urine was spiked with the standard solutions of the mercapturic acids at concentrations of 40 or 400 µg/L, divided into six aliquots and analysed. The relative recovery rates obtained are given in Table 9.

Table 9 Mean relative recovery rates for mercapturic acids in pooled urine (n = 6).

Analyte	Spiked analyte level [µg/L]	Mean rel. recovery [%]	Range [%]
AAMA	40	110	96–122
	400	100	95–102
HEMA	40	97	89–107
	400	103	99–110
3-HPMA	40	122	109–141
	400	131	127–137
2-HPMA	40	109	95–120
	400	104	95–112
AMCC	40	117	98–131
	400	117	111–124

As, under the given chromatographic conditions, the analytes (especially 3-HPMA and AMCC) elute in part simultaneously and yield very similar mass specific transitions, a further test for checking specifity was carried out. For this purpose, an unspiked individual urine sample was prepared according to Section 3 and analysed according to Section 4 (blank value). Subsequently, the same urine samples were analysed for another four times, each time spiked with 1000 µg/L of one of the analytes AAMA, HEMA, 3-HPMA or AMCC. This was done to investigate the influence of individual high analyte levels on the other analytes. The results of this investigation are summarised in Table 10.

Table 10 Test results for specifity of the individual analytes.

Spiked analyte [1 mg/L]	Conc. AAMA [µg/L]	Conc. HEMA [µg/L]	Conc. 3-HPMA [µg/L]	Conc. AMCC [µg/L]
Blank value	103.4	3.0	153.5	216.2
AAMA	1188.6	3.1	145.1	209.8
HEMA	131.6	991	163.4	206.1
3-HPMA	101.8	2.6	1208.7	221.2
AMCC	129.6	3.3	180.2	1356.7

9.3 Absolute recovery

Additionally, by the examiner of the method, the determination of absolute recovery or analyte losses due to solid phase extraction was carried out. For this purpose, pooled urine was spiked in two different concentration levels with the analytes before and after solid phase extraction and the peak areas obtained were compared. The resulting absolute recovery rates are shown in Table 11.

Table 11 Absolute recovery rates of the analytes after solid phase extraction.

Analyte	Spiked level 1 [µg/L]	Abs. recovery	Spiked level 2 [µg/L]	Abs. recovery
HEMA	45	45.1%	400	43.0%
AAMA	116	61.8%	479	56.9%
2-HPMA	33	45.7%	360	53.5%
3-HPMA	132	56.0%	599	47.6%
AMCC	185	97.1%	564	95.4%

9.4 Matrix effects

The matrix effects, i.e. the influence of the presence of matrix on signal intensity, was also determined by the examiner of the method. The matrix effects (ME) were determined by spiking the eluate of three different urine samples and two blank samples (water) with IS and the analytes. Then, the spiked urine samples and the spiked water samples were compared with regard to the peak areas of the analytes and the internal standards (taking the background level of each analyte into account). Determination of the background level of the analytes was carried out three times by preparing unspiked urine samples according to Section 3.

An ME of 100% demonstrates that the presence of matrix has no influence on the signal intensities of the substances. An ME of less than 100% means a suppression of the analyte signal, an ME greater than 100% means an improvement of the signal via matrix. The data obtained are listed in Table 12.

Table 12   Determined matrix effects of analytes and of the respective internal standards.

Analyte	Analyte level 1 [µg/L]	Matrix effect	Analyte level 2 [µg/L]	Matrix effect
HEMA	40	~22%	400	~18%
AAMA	40	~10%	400	~13%
2-HPMA	40	~32%	400	~34%
3-HPMA	40	~33%	400	~36%
AMCC	40	~64%	400	~40%

9.5   Stability of the analytes

To investigate the stability of the analytes after a number of deep-freeze/thawing cycles, three quality control samples (each spiked with 400 µg/L each of the analytes) were subjected to deep freezing (at –20°C) and rethawing on three days and were subsequently analysed.

The deviation from the mean values of the quality control samples was within the range of measurement uncertainty. Degradation of the analytes by repeated freezing and thawing was not observed.

9.6   Detection limits and quantitation limits

Under the given conditions the detection and quantitation limits shown in Table 13 were obtained. The detection limit was estimated on the basis of a signal to noise ratio of 3:1. The quantitation limit was calculated on the basis of a signal to noise ratio of 9:1.

Table 13   Detection and quantitation limits of the mercapturic acids in urine.

Analyte	Detection limit [µg/L]	Quantitation limit [µg/L]
AAMA	5.0	15
HEMA	0.5	1.5
3-HPMA	1.0	3.0
AMCC	5.0	15
2-HPMA	1.0	3.0

9.7   Sources of error

Due to the matrix of the sample, ionisation of the analytes can be suppressed to a more or less distinct extent (so-called quenching). This is noticeable in rare cases where samples contain high protein contents, and results in a higher detection limit than given here. As a rule, the chromatograms are free of noteworthy interference peaks. Only for AAMA an interference peak elutes just behind the analyte peak. However, these peaks can be easily separated by a perpendicular drop line. Here, in order to exclude possible false positive values, careful attention must be paid to the peak form. In the case of re-injections, deviations in analyte concentration levels have not been observed up to now. Also, carry-over contamination effects as well as reagent blank values were not observed.

The solid phase extraction steps have to be performed accurately according to the instructions in Section 3. The highly polar mercapturic acids can easily be rinsed off the column if the wrong solutions or diverging solution quantities are applied. It has to be pointed out again that all solid phase extraction steps should be carried out without the help of a vacuum pump, if possible.

The introduction of an isotopically labelled internal standard for 2-HPMA ($^{13}C_2$-2-HPMA) has significantly improved the precision and accuracy for this parameter. Nevertheless, a minor disadvantage is that the linear range of the calibration only reaches as far as 500 µg/L urine and urine samples with higher 2-HPMA levels have to be diluted and prepared again. This can be attributed to mass spectrometric interferences of the analyte with the internal standard as the IS for 2-HPMA is labelled only twice. Due to the natural occurrence of ^{13}C (1.1%), overlapping of the analysed mass fragments occurs when using a mass spectrometer with a resolution of $m/z = +/-0.5\ m/z$. This results in an increase of the peak area of the internal standard (and thus to a flattening of the calibration graph) when high concentrations of the analyte are found. However, even in the urine of smokers, the background exposure level of 2-HPMA is clearly below 500 µg/L (see Table 1).

10 Discussion of the method

The method permits simultaneous, sensitive and highly specific quantitation of mercapturic acids as metabolites of important industrial chemicals such as acrylamide, ethylene oxide, propylene oxide, dimethylformamide and acrolein in the urine of individuals from the general population as well as in urine samples of workers occupationally exposed to these substances.

One decisive advantage of this method is the simultaneous quantitation of the mercapturic acids, whereby the method is excellently suitable for screening investigations. The procedure is based on the method published by Böttcher and Angerer to determine the mercapturic acids of acrylamide and glycidamide [5] and was extended to its present form to include the other mercapturic acids. As it is difficult to enrich the highly polar mercapturic acids specifically on the solid phases used, the analysis is characterised by extracts with a relatively high matrix content. Therefore, to protect the LC-MS/MS system, the use of a switch valve is recommended, in order to conduct the flow of the analytical column into the MS/MS system within a limited time window only. Contrary to the original method, isocratic elution of the mercapturic acids was replaced by a gradient elution, which allows rinsing the analytical column to a certain extent in order to exclude memory effects. Composition of the eluents has been optimised to achieve a gradient as straight and as reproducible as possible with the necessary low flow rates.

To avoid the injection of large sample quantities into the LC-MS/MS system, as an alternative, the residue of the eluate can also be dissolved in 100 µL 0.1% formic acid (instead of 500 µL), of which then only 20 µL (instead of 100 µL) are injected.

As the good relative recovery rates as well as the very good precision data show, the matrix effects as well as the analyte losses occurring through solid phase extraction can effectively be compensated using isotopically labelled internal standards. As confirmation for the minor importance of matrix effects on the accuracy of the method, the comparison between the calibration graphs in water or urine can be consulted, which shows only minor deviations.

As the validation data show, the method presented is very specific for the different mercapturic acids, and a mutual influence between the individual analytes was not observed. The detection limits obtained are adequate for the detection of mercapturic acids in the urine of the general population.

The presented method is very well suited for monitoring the internal exposure of individuals environmentally or occupationally exposed to acrylamide, ethylene oxide, propylene oxide, acrolein or dimethylformamide.

Instruments used:

- HPLC 1100 series of Agilent with an integrated autosampler as well as tandem mass spectrometric detector API-3000 by Applied Biosystems, evaluation software: Analyst Version 1.3, Applied Biosystems.

11 References

[1] A. Hartwig (ed.): Acrylamide. The MAK Collection Part I: MAK Value Documentations. Volume 25, Wiley-VCH, Weinheim (2009).

[2] T. Schettgen: Biochemisches Effekt-Monitoring in der Umweltmedizin – Hämoglobin-Addukte von Acrylamid, Glycidamid und Acrylnitril im Blut der Allgemeinbevölkerung. Dissertation an der Naturwissenschaftlichen Fakultät der Universität Erlangen-Nürnberg (2006).

[3] E. Tareke, P. Rydberg, P. Karlsson, S. Eriksson, M. Törnqvist: Analysis of acrylamide, a carcinogen formed in heated foodstuffs. J. Agric. Food Chem. 50, 4998–5006 (2002).

[4] S.C. Sumner, C.C. Williams, R.W. Snyder, W.L. Krol, B. Asgharian, T.R. Fennell: Acrylamide: a comparison of metabolism and hemoglobin adducts in rodents following dermal, intraperitoneal, oral, or inhalation exposure. Toxicol. Sci. 75, 260–270 (2003).

[5] M.I. Böttcher and J. Angerer: Determination of the major mercapturic acids of acrylamide and glycidamide in human urine by LC-ESI-MS/MS. J. Chromatogr. B 824, 283–294 (2005).

[6] M.I. Böttcher, H.M. Bolt, H. Drexler, J. Angerer: Excretion of mercapturic acids of acrylamide and glycidamide after single oral administration of deuterium-labelled acrylamide. Arch. Toxicol. 80, 55–61 (2006).

7 Deutsche Forschungsgemeinschaft (DFG): List of MAK and BAT values 2012, 48th issue Wiley-VCH, Weinheim (2012).
8 Ausschuss für Gefahrstoffe (AGS): Bekanntmachung zu Gefahrstoffen, Bekanntmachung 910: Risikowerte und Expositions-Risiko-Beziehungen für Tätigkeiten mit krebserzeugenden Gefahrstoffen. Zuletzt geändert: GMBl 2011, S. 194 [Nr. 10].
9 H. Greim (ed.): Ethylene oxide. MAK Value Documentations. Volume 5, Wiley-VCH, Weinheim (1993).
10 H. Greim (ed.): Ethylenoxid. Gesundheitsschädliche Arbeitsstoffe, Toxikologisch-arbeitsmedizinische Begründung von MAK-Werten, 34th issue, Wiley-VCH, Weinheim (2002).
11 J. Angerer, M. Bader, A. Krämer: Ambient and biochemical effect monitoring of workers exposed to ethylene oxide. Int. Arch. Occup. Environ. Health 71, 14–18 (1998).
12 T. Schettgen, H.C. Broding, J. Angerer, H. Drexler: Hemoglobin adducts of ethylene oxide, propylene oxide, acrylonitrile and acrylamide – biomarkers in occupational and environmental medicine. Toxicol. Letters 134, 65–70 (2002).
13 M. Gérin and R. Tardif: Urinary N-acetyl-S-2-hydroxyethyl-cysteine in rats as biological indicator of ethylene oxide exposure. Fundam. Appl. Toxicol. 7, 419–423 (1986).
14 M. Müller, A. Krämer, J. Angerer, E. Hallier: Ethylene oxide-protein adduct formation in humans: influence of glutathione-S-transferase polymorphisms. Int. Arch. Occup. Environ. Health 71, 499–502 (1998).
15 J.F. Stevens and C.S. Maier: Acrolein: sources, metabolism, and biomolecular interactions relevant to human health and disease. Mol. Nutr. Food Res. 52, 7–25 (2008).
16 H. Greim (ed.): Acrolein. MAK Value Documentations. Volume 13, Wiley-VCH, Weinheim (2001).
17 H. Greim (ed.): Propylene oxide. MAK Value Documentations. Volume 5, Wiley-VCH, Weinheim (1993).
18 P.J. Boogaard, P.S.J. Rocchi, N.J. van Sittert: Biomonitoring of exposure to ethylene oxide and propylene oxide by determination of haemoglobin adducts: correlation between airborne exposure and adduct levels. Int. Arch. Occup. Environ. Health 72, 142–150 (1999).
19 J. Mráz and H. Nohová: Absorption, metabolism and elimination of N,N-dimethylformamide in humans. Int. Arch. Occup. Environ. Health 64, 85–92 (1992).
20 H. Greim (ed.): Dimethylformamide. MAK Value Documentations. Volume 8, Wiley-VCH Verlag, Weinheim (1997).
21 A. Hartwig (ed.): Dimethylformamide. The MAK Collection Part I: MAK Value Documentations. Volume 26, Wiley-VCH Verlag, Weinheim (2010).
22 T. Schettgen, A. Musiol, T. Kraus: Simultaneous determination of mercapturic acids derived from ethylene oxide (HEMA), propylene oxide (2-HPMA), acrolein (3-HPMA), acrylamide (AAMA) and N,N-dimethylformamide (AMCC) in human urine using liquid chromatography/tandem mass spectrometry. Rapid Comm. Mass Spectrom. 22, 2629–2638 (2008).
23 M. Kellert, K. Scholz, S. Wagner, W. Dekant, W. Völkel: Quantitation of mercapturic acids from acrylamide and glycidamide in human urine using a column switching tool with two trap columns and electrospray tandem mass spectrometry. J. Chromatogr. A 1131, 58–66 (2006).
24 A.M. Calafat, D.B. Barr, J.L. Pirkle, D.L. Ashley: Reference range concentrations of N-acetyl-S-(2-hydroxyethyl)-L-cysteine, a common metabolite of several volatile organic compounds, in the urine of adults in the United States. J. Exp. Anal. Environ. Epidemiol. 9, 336–342 (1999).
25 D.G. Mascher, H.J. Mascher, G. Scherer, E.R. Schmid: High-performance liquid chromatographic-tandem mass spectrometric determination of 3-hydroxypropyl-mercapturic acid in human urine. J. Chromatogr. B 750,163–169 (2001).

26 H.U. Käfferlein and J. Angerer: Determination of N-Acetyl-S-(N-methylcarbamoyl)-cysteine (AMCC) in the general population using gas-chromatography-mass spectrometry. J. Environ. Monit. 1, 465–469 (1999).

27 Bundesärztekammer: Richtlinie der Bundesärztekammer zur Qualitätssicherung quantitativer laboratoriumsmedizinischer Untersuchungen. Dt. Ärztebl. 100, A3335–A3338 (2003).

28 J. Angerer, T. Göen, G. Lehnert: Mindestanforderungen an die Qualität von umweltmedizinisch-toxikologischen Analysen. Umweltmed. Forsch. Prax. 3, 307–312 (1998).

Author: *Th. Schettgen*
Examiners: *G. Scherer, K. Sterz*

12 Appendix

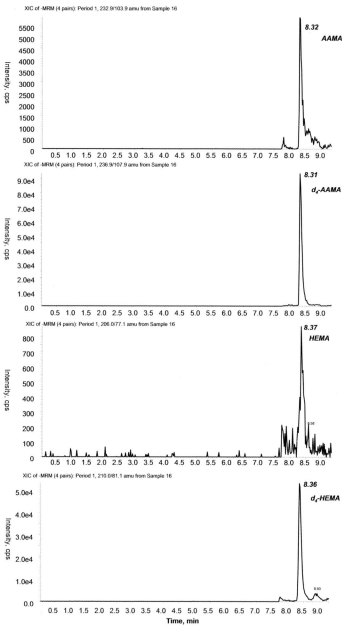

Fig. 5 Chromatogram of a prepared urine sample from a non-smoker (creatinine: 0.68 g/L) with a determined concentration level of 24 μg/L AAMA and 1.7 μg/L HEMA.

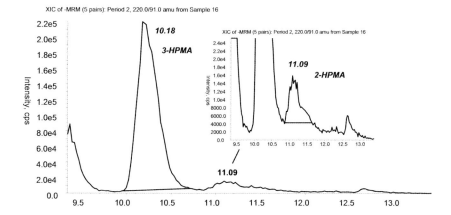
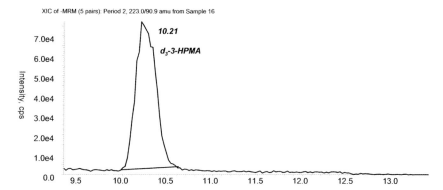
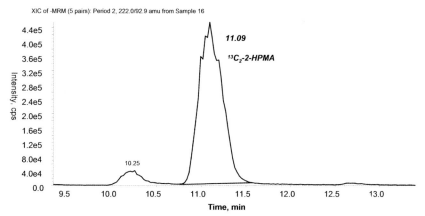

Fig. 6 Chromatogram of a prepared urine sample from a non-smoker (creatinine: 0.68 g/L) with a determined concentration level of 389 µg/L 3-HPMA and 12.9 µg/L 2-HPMA.

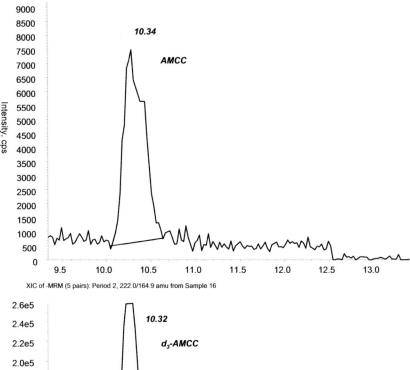
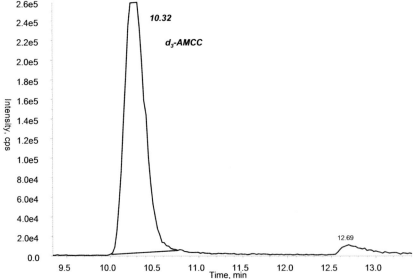

Fig. 7 Chromatogram of a prepared urine sample from a non-smoker (creatinine: 0.68 g/L) with a determined concentration level of 13.4 μg/L AMCC.

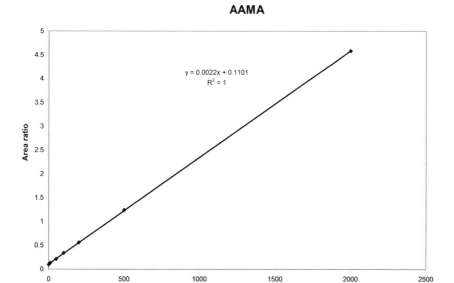

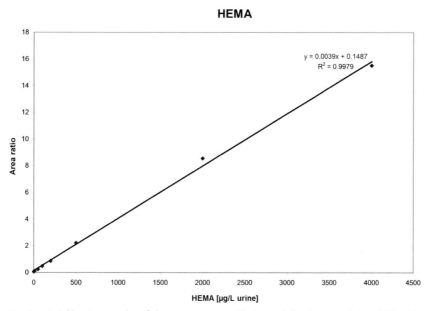

Fig. 8 a) Calibration graphs of the mercapturic acids in pooled urine (creatinine: 0.52 g/L): AAMA and HEMA.

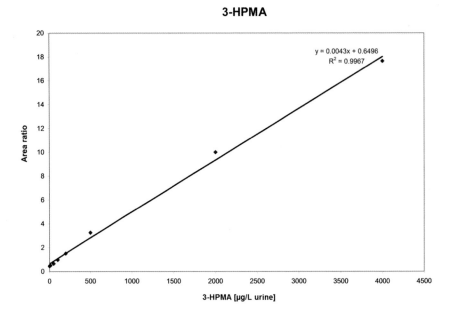

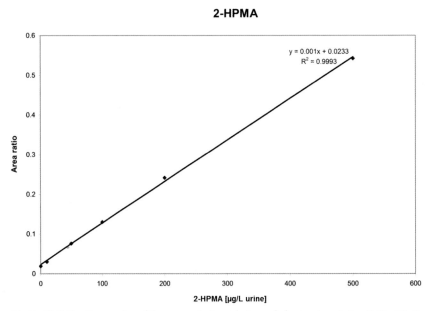

Fig. 8 b) Calibration graphs of the mercapturic acids in pooled urine (creatinine: 0.52 g/L): 3-HPMA and 2-HPMA.

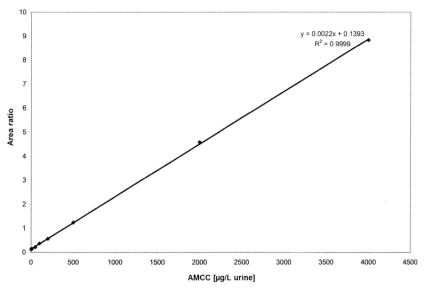

Fig. 8 c) Calibration graphs of the mercapturic acids in pooled urine (creatinine: 0.52 g/L): AMCC.

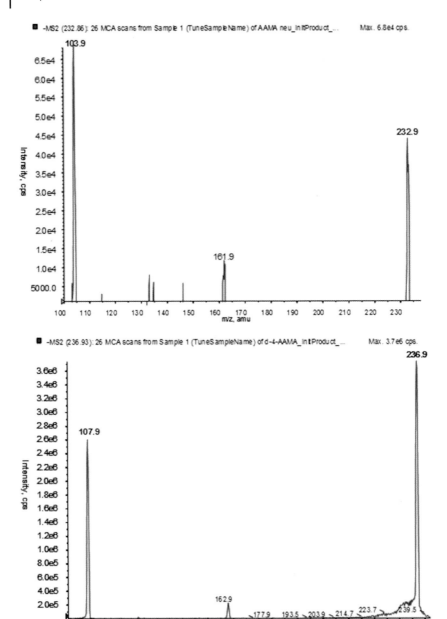

Fig. 9 a) ESI negative spectra of the product ion scans for AAMA and d_4-AAMA.

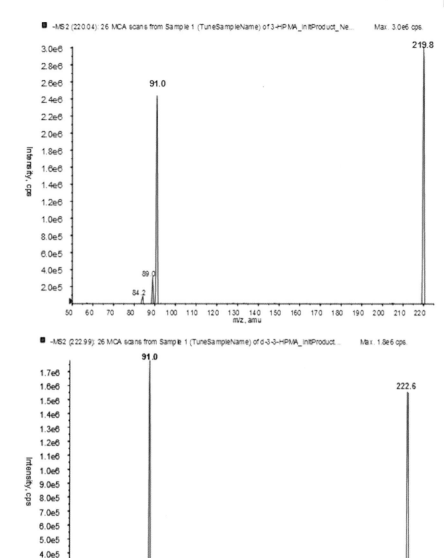

Fig. 9 b) ESI negative spectra of the product ion scans for 3-HPMA and d_3-3-HPMA.

Analytical Methods

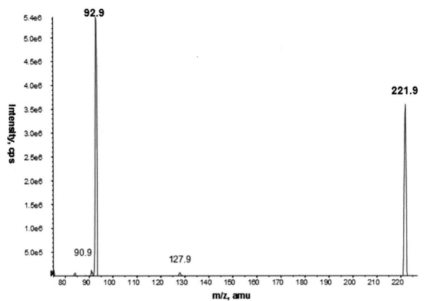

Fig. 9   c) ESI negative spectrum of the product ion scan for $^{13}C_2$-2-HPMA.

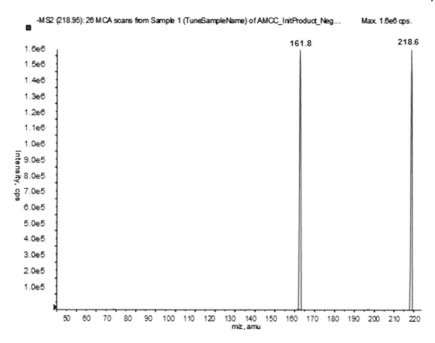
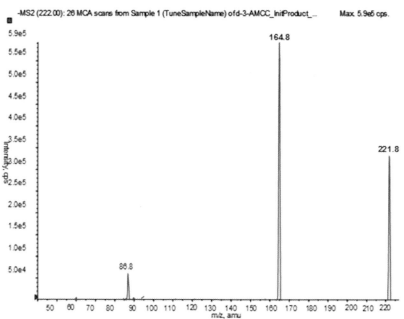

Fig. 9 d) ESI negative spectra of the product ion scans for AMCC and d_3-AMCC.

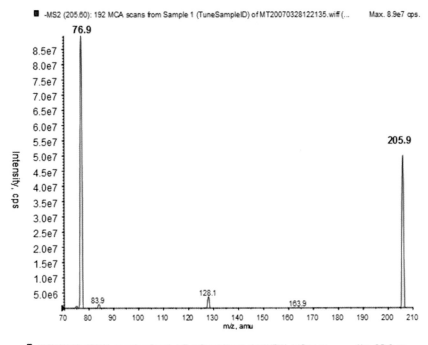
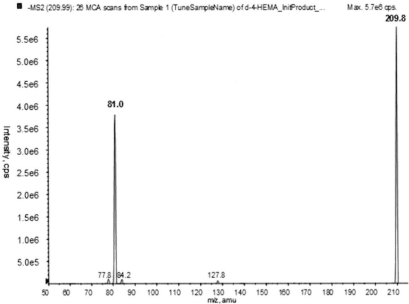

Fig. 9   e) ESI negative spectra of the product ion scans for HEMA and d_4-HEMA.

2-Phenyl-2-propanol in urine

Matrix:	Urine
Hazardous substance:	Isopropyl benzene (cumene)
Analytical principle:	Gas chromatography/flame ionisation detection (GC/FID)
Completed in:	April 2002

Overview of parameters that can be determined with this method and the corresponding chemical substances:

Hazardous substance	CAS	Parameter	CAS
Isopropyl benzene (cumene)	98-82-8	2-Phenyl-2-propanol	617-94-7

Summary

2-Phenyl-2-propanol is the main metabolite of isopropyl benzene (cumene) and can therefore be used as a biomarker for exposure to cumene. The procedure described here permits the determination of 2-phenyl-2-propanol in urine. Relevant occupational exposure to cumene can thus be determined.

After addition of the internal standard 2-(4-fluorophenyl)ethanol the urine is first subjected to hydrolysis using hydrochloric acid. The analyte is then enriched using liquid-liquid extraction and simultaneously separated from matrix components. After separation by capillary gas chromatography, the extract is measured using a flame ionisation detector.

Calibration curves are plotted for quantitative evaluation using calibration standard solutions prepared by spiking pooled human urine with standard substance. The calibration standards are treated in the same manner as the urine samples to be investigated.

2-Phenyl-2-propanol

Within day precision:	Standard deviation (rel.)	s_w = 8.6% or 6.5%
	Prognostic range	u = 19.2% or 14.5%
	at concentrations of 5 mg or 48 mg 2-phenyl-2-propanol per litre urine and where n = 10 determinations	
Day to day precision:	Standard deviation (rel.)	s_w = 9.8% or 7.5%
	Prognostic range	u = 21.4% or 16.3%
	at concentrations of 5 mg or 48 mg 2-phenyl-2-propanol per litre urine and where n = 12 determinations	
Accuracy:	Recovery rate	r = 109% or 91.9%
	at concentrations of 1mg or 100 mg 2-phenyl-2-propanol per litre urine and where n = 5 determinations	
Detection limit:	0.5 mg 2-phenyl-2-propanol per litre urine	
Quantitation limit:	1.5 mg 2-phenyl-2-propanol per litre urine	

Isopropyl benzene

Isopropyl benzene (cumene) is a colourless aromatic hydrocarbon insoluble in water with an odour similar to that of petrol/gasoline. Cumene is unlimitedly miscible with other solvents as acetone, alcohol, ether or benzene. With air its vapours form an explosive mixture. The approx. 800,000 tons per year produced at present in Germany almost exclusively involve the production of phenol, acetone and α-methylstyrene. Furthermore, cumene is used as basic material for the industrial production of detergents and other chemical products, as plastics, resins, pesticides and insecticides. It is also used as a metal cleaning agent and as a solvent for paints, printing inks, inks and lacquers. It is furthermore used as an additive in carburetting fuels [1–3].

Aromatic hydrocarbons are local irritants. The toxicity of cumene is comprehensively described in the 1996 MAK value documentation [4].

When handling cumene, inhalation, dermal and oral absorption is in principle possible. From occupational and environmental medical viewpoint main uptake occurs via percutaneous absorption and especially via inhalation. In experimental inhalation studies, the pulmonary retention in humans was non-concentration-dependent at an initial 64% and decreased after 8-hour exposure to 45% [5, 6]. No studies on the distribution and accumulation of cumene in human tissues are available at present. Partition coefficients of water, blood and oil with air are 1.4, 37 and 6215, respectively, of oil / water > 4300 and of oil / blood 168 leading to the conclusion that cumene is distributed throughout the entire organism – especially in the fatty tissue [7, 8]. The biotransformation of cumene occurs by stepwise oxidation, mainly in the liver [9, 10]. The oxidative metabolism of cumene starts at the isopropyl group. The biotransformation scheme is given in Figure 1.

The spectrum of metabolites consists primarily of 2-phenyl-2-propanol and, in addition, of 2-phenyl-1-propanol and 2-phenylpropionic acid. The metabolites are bound to glucuronic acid and glycine, respectively [11]. Unlike other aromatic hy-

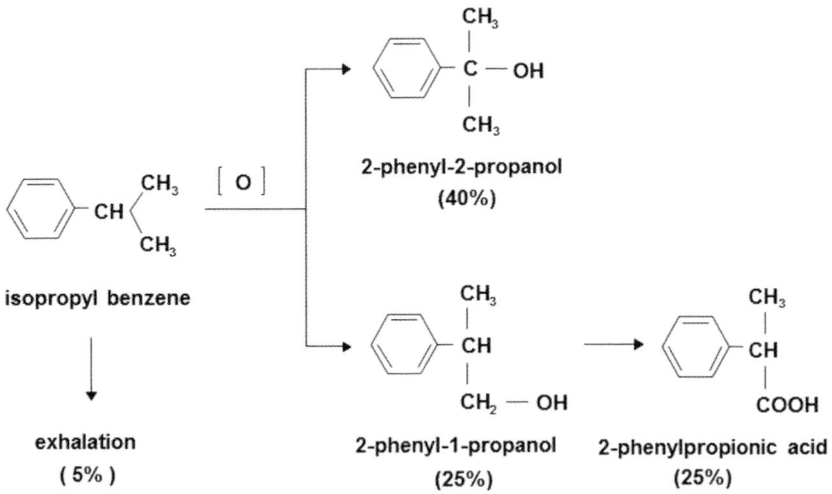

Fig. 1 The metabolism of cumene. The elimination rates are based on animal studies [6, 9].

drocarbons, no degradation via phenol derivatives takes place, a fact to which the absence of toxicity to the bone marrow is attributed [5, 6, 9, 11]. According to inhalation studies in humans, 35% of an inhaled cumene dose is eliminated biphasically as 2-phenyl-2-propanol [6]. This metabolite should thus be preferred as biomonitoring parameter in so far as it has the highest elimination rate.

Results of field studies on the relationship between external and internal cumene exposure are not available at present. Knecht et al. conducted a laboratory study with 18 volunteers with an external cumene exposure between 15 and 50 ppm [12]. Their investigations involving 4-hour exposure to 50 ppm cumene revealed an initial pulmonary retention of 50.1%, which decreased to 37.8% at the end of exposure. It could also be shown that 50% of the main metabolite 2-phenyl-2-propanol is eliminated about 9.5 hours after initial exposure. No pronounced cumulative effect can be derived from this elimination kinetics. Independently of external exposure, the calculated average half-life was 6.4 ± 1.0 hours. Based on the laboratory investigations of Knecht et al. [12], an association between the cumene concentration in the ambient air and the urinary 2-phenyl-2-propanol concentration can be established. Accordingly, 8-hour exposure to cumene at a level of 50 ppm corresponds to renal excretion of about 50 mg 2-phenyl-2-propanol/g creatinine at the end of exposure. The BAT value (*Biologischer Arbeitsstoff-Toleranzwert*) was established at 50 mg 2-phenyl-2-propanol/g creatinine (after hydrolysis) [13].

Author: *U. Knecht*
Examiners: *Th. Göen, M. Müller*

2-Phenyl-2-propanol in urine

Matrix:	Urine
Hazardous substance:	Isopropyl benzene (cumene)
Analytical principle:	Gas chromatography/flame ionisation detection (GC/FID)
Completed in:	April 2002

Contents

1	General principles
2	Equipment, chemicals and solutions
2.1	Equipment
2.2	Chemicals
2.3	Solutions
2.4	Internal standard
2.5	Calibration standards
3	Specimen collection and sample preparation
4	Operational parameters
5	Analytical determination
6	Calibration
7	Calculation of the analytical result
8	Standardisation and quality control
9	Evaluation of the method
9.1	Precision
9.2	Accuracy
9.3	Detection limit and quantitation limit
9.4	Sources of error
10	Discussion of the method
11	References
12	Appendix

1 General principles

The acidified urine samples are hydrolysed together with the internal standard 2-(4-fluorophenyl)ethanol, then neutralised with sodium hydroxide, saturated with sodium chloride and extracted twice with diethylether. After addition of n-butylacetate as keeper the ether extracts are carefully evaporated. Derivatisation of the residue is carried out at 75°C with pyridine and acetic anhydride. Detection is performed by gas chromatographic separation using flame ionisation detection. Calibration is performed using calibration standard solutions prepared by spiking pooled human urine with known 2-phenyl-2-propanol concentrations. The calibration standards are treated in the same manner as the urine samples to be investigated.

2 Equipment, chemicals and solutions

2.1 Equipment

- Gas chromatograph with split-/splitless injector, flame ionisation detector (FID), autosampler and data processing system
- Capillary gas chromatographic column: length: 50 m; inner diameter: 0.53 mm; film thickness: 2 µm, stationary phase: 14%-cyanopropylphenyl-86%-dimethylpolysiloxane; (e.g. CP-Sil 19 CB of Agilent Technologies)
- 250 mL Urine containers with screw cap (e.g. Sarstedt No. 77.577)
- Analytical balance (e.g. Sartorius)
- 20 mL Crimp cap vials with PTFE (polytetrafluoroethylene) septa and crimp caps as well as a manual crimper (e.g. Macherey-Nagel)
- Thermostatic water bath
- 10 mL, 50 mL and 100 mL volumetric flasks
- Piston-stroke pipettes with adjustable volume between 10–100 µL or 100–1000 µL with suitable pipette tips (e.g. Eppendorf)
- Finnpipette 1000–5000 µL (e.g. Thermo Scientific)
- Laboratory centrifuge
- 2 mL screw cap vials with insert and screw caps with PTFE septa
- Laboratory shaker
- Drying oven
- Evaporator for sample concentration

2.2 Chemicals

- Acetone p.a. (e.g. Merck, No. 100012)
- Hydrochloric acid 37% p.a. (e.g. Merck, No. 100317)
- 2-Phenyl-2-propanol 97% (e.g. Aldrich, No. P30802)

- 2-(4-Fluorophenyl)ethanol p.a. (e.g. Fluka, No. 47285)
- Sodium hydroxide pellets p.a. (e.g. Merck, No. 106498)
- Sodium chloride p.a. (e.g. Merck, No. 106404)
- Diethylether p.a. (e.g. Merck, No. 100921)
- Pyridine p.a. (e.g. Merck, No. 109728)
- Acetic anhydride p.a. (e.g. Merck, No. 100042)
- n-Butylacetate p.a. (e.g. Merck, No. 109652)
- Bidist. water
- Nitrogen 5.0 (e.g. Linde)
- Helium 5.0 (e.g. Linde)

2.3 Solutions

- Sodium hydroxide solution (12 mol/L)
  60 mL bidist. water are placed in a glass beaker. Subsequently, under stirring, 48.0 g sodium hydroxide are added. The solution is transferred to a 100 mL volumetric flask and then made up to the mark with bidist. water. The concentration of the sodium hydroxide solution is 480 g/L.

2.4 Internal standard

- Stock solution
  100 mg of the 2-(4-fluorophenyl)ethanol are weighed into a 10 mL volumetric flask. The flask is then made up to the mark with acetone. The concentration of the internal standard in the stock solution is 10 g/L.
- Working solution:
  1 mL stock solution is pipetted into a 50 mL volumetric flask. The flask is then made up to the mark with acetone/bidist. water (1:1). The concentration of the internal standard in the working solution is 200 mg/L. 200 µL each are added to samples, calibration standards and quality controls.

2.5 Calibration standards

- Working solution
  100 mg 2-phenyl-2-propanol are weighed into a 10 mL volumetric flask. The flask is then made up to the mark with acetone. The concentration of 2-phenyl-2-propanol in the working solution is 10 g/L.

From this working solution calibration standards in urine are prepared by pipetting between 100 µL and 2.0 mL of the working solution into 100 mL volumetric flasks and making them up to the mark with pooled urine. Concentrations between 10 mg and 200 mg 2-phenyl-2-propanol per litre urine are obtained.

Table 1 Pipetting scheme for the preparation of 2-phenyl-2-propanol calibration standards in urine.

Volume of the working solution	Final volume of the calibration standard solution	Analyte level 2-phenyl-2-propanol
[mL]	[mL]	[mg/L]
2.0	100	200
1.5	100	150
1.0	100	100
0.50	100	50
0.25	100	25
0.10	100	10

The calibration standards can be kept in the refrigerator at −18°C for at least 6 months.

3 Specimen collection and sample preparation

The urine samples are collected in sealable plastic containers and stored in the refrigerator at −18°C until analysis is performed. Under these storage conditions, the urine is stable for at least one year.

The urine samples are defrosted in the water bath at 60°C. After equilibration to room temperature and vigorous shaking, 2 mL urine are pipetted into a 20 mL crimp cap vial and 200 µL of the working solution of the internal standard (200 mg/L) are added.

After acidification of the sample with 200 µL 37% hydrochloric acid, the crimp cap vial is sealed and hydrolysis is performed for one hour at 90°C in the water bath. After cooling to room temperature the hydrolysate is neutralised with 400 µL sodium hydroxide (12 mol/L) and saturated with 0.6 g sodium chloride. After the addition of 2 mL diethylether the crimp cap vial is sealed again and thoroughly shaken for 10 min. It is then centrifuged for 10 min at 4000 U/min. The ether phase is removed and the extraction step repeated a second time. 100 µL n-butyl-acetate are added to the combined ether extracts as keeper and the ether is then carefully removed under a stream of nitrogen. 100 µL pyridine as well as 100 µL acetic anhydride are added to the residue (approx. 100 µL) for derivatisation of the 2-phenyl-2-propanol as well as of the internal standard and heated for one hour at 75°C in the drying oven. The residue is transferred to an insert and analysed using GC-FID.

4 Operational parameters

Capillary column:	Material:	Fused Silica
	Stationary phase:	CP-SiI 19 CB
	Length:	50 m
	Inner diameter:	0.53 mm
	Film thickness:	2.0 µm
Detector:	Flame ionisation detector (FID)	
Temperature:	Column:	Initial temperature 175°C; 12 min isothermal, then increase at a rate of 50°C/min to 270°C; 5 min isothermal at final temperature
	Injector:	220°C
	Detector:	280°C
Carrier gas:	Helium 5.0 at a column pre-pressure of 50 kPa	
Split:	1:10	
Injection volume:	1 µL	

All other parameters must be optimised in accordance with the manufacturer's instructions.

5 Analytical determination

For gas chromatographic analysis of the urine samples processed according to Section 3, 1 µL of the residue is injected into the GC/FID system. Figure 2 (in the Appendix) shows as an example the chromatogram of a urine sample from a volunteer who was exposed to cumene. The retention times of the 2-phenyl-2-propanol as well as of the internal standard shown in Figure 2 are intended to be a rough guide only. Users of the method must ensure proper separation performance of the analytical column they use and of the resulting retention behaviour of the analyte.

A reagent blank value is included within each analytical series. Here, instead of urine bidist. water is subjected to the described sample preparation.

6 Calibration

The calibration standards prepared in urine according to Section 2.5 are processed in the same manner as the urine samples (see Section 3) and analysed using GC/FID. The calibration graphs are obtained by plotting the quotient of the peak area of 2-phenyl-2-propanol and of the internal standard against the concentration used. It is not necessary to perform a complete calibration for every analytical series. As a general rule, the analysis of one calibration standard is mostly sufficient. In this case, the value obtained for this standard is related to the one from the complete calibration curve. A new calibration should be performed when the results of quality control indicate systematic deviations.

The calibration curve is linear between 1 mg and 200 mg 2-phenyl-2-propanol per litre urine. In Figure 3 (in the Appendix) an example of a calibration curve is shown.

7 Calculation of the analytical result

The analyte concentration in urine samples is calculated on the basis of the plotted calibration graph (see Section 6). The peak area of the respective analyte is divided by the peak area of the internal standard. From the quotient thus obtained, the corresponding concentration of the analyte in mg per litre urine can be determined from the relevant calibration graph. A possible reagent blank value has to be taken into consideration.

8 Standardisation and quality control

Quality control of the analytical results is carried out as stipulated in the guidelines of the *Bundesärztekammer* (German Medical Association) [14]. For quality control, urine spiked with a defined amount of 2-phenyl-2-propanol is analysed within every analytical series. A quality control material for 2-phenyl-2-propanol in urine is not commercially available. Thus, it has to be prepared in the laboratory. For this purpose the urine of individuals occupationally non-exposed to cumene is spiked with a defined quantity of 2-phenyl-2-propanol in the relevant concentration range. The control urine is divided into aliquots and can be stored for up to twelve months when deep frozen at approx. −20°C. The nominal value and the tolerance ranges of the quality control material are determined in a pre-analytical period (one analysis of the control material on each of 15 different days) [15, 16].

9 Evaluation of the method

9.1 Precision

To determine within day precision, defined quantities of 2-phenyl-2-propanol are spiked to pooled urine of persons occupationally non-exposed to cumene to obtain concentrations of 5 mg and 25 mg/L as a result. The urine sample of a volunteer exposed to cumene (50 ppm/m^3 air) with an expected value of 48 mg/L was also included in the determination of the precision data. For within day precision ten replicates of the urine samples yielded the following relative standard deviations with the corresponding prognostic ranges (Table 2).

Table 2 Within day precision for the determination of 2-phenyl-2-propanol in urine (n = 10).

Analyte	Spiked concentration [mg/L]	Standard deviation (rel.) s_w [%]	Prognostic range u [%]
2-Phenyl-2-propanol	5	8.6	19.2
	25	7.1	15.8
	48	6.5	14.5

Furthermore, the precision from day to day was determined. For this purpose, the spiked urine was processed on twelve different days and analysed as described in the preceding sections. The obtained precision data are shown in Table 3.

Table 3 Precision from day to day for the determination of 2-phenyl-2-propanol in urine (n = 12).

Analyte	Spiked concentration [mg/L]	Standard deviation (rel.) s_w [%]	Prognostic range u [%]
2-Phenyl-2-propanol	5	9.8	21.4
	25	8.1	17.7
	48	7.5	16.3

9.2 Accuracy

To determine the accuracy of the method recovery experiments were carried out. For this purpose defined quantities of 2-phenyl-2-propanol are added to the pooled urine of persons occupationally non-exposed to cumene leading to concentrations from 1 mg/L up to 100 mg/L. These samples were processed and analysed from five to nine times corresponding to Section 3. Mean recovery rates between 91.9% and 109.3% (Table 4) were obtained.

Table 4 Relative recovery rates for the determination of 2-phenyl-2-propanol in urine (n = 5–9).

Analyte	Nominal concentration [mg/L]	Recovery rate r [%]	Range [%]
2-Phenyl-2-propanol	1	109.3	80.0–140.0
	25	92.8	77.2–108.4
	50	103.3	98.2–108.4
	75	104.2	99.47–108.9
	100	91.9	87.3–96.5

Analyte losses due to sample preparation were determined by the examiner of the method. For this purpose, the pooled urine of individuals occupationally non-exposed to cumene was prepared with three different 2-phenyl-2-propanol concentrations and divided into aliquots. The urine samples were processed five times and the measured concentrations compared with samples processed without hydrolysis. The absolute recovery rates derived lie within the range of 48.2% to 55.0% (Table 5).

Table 5 Absolute recovery rates for 2-phenyl-2-propanol in urine (n = 5).

Analyte	Nominal concentration [mg/L]	Absolute recovery rate [%]	Range [%]
2-Phenyl-2-propanol	5	48.3	26.4–66.0
	25	48.2	40.6–62.6
	50	55.0	49.1–60.8

9.3 Detection limit and quantitation limit

The detection limit was about 0.5 mg 2-phenyl-2-propanol per litre urine (three times the signal/background to noise ratio) under the conditions given for sample preparation and determination by means of GC-FID. This corresponds to a quantitation limit of 1.5 mg 2-phenyl-2-propanol per litre urine.

9.4 Sources of error

Using acid hydrolysis of the 2-phenyl-2-propanol bound to glucuronic acid or glycin described here, a decomposition of the 2-phenyl-2-propanol, which is unstable in an acidic environment, takes place as a result of dehydration reactions at the side chain of the molecule. The absolute recovery rates are on average between 48% and 55% (Table 5).

Enzymatic hydrolysis of the conjugates is also possible. However, compared with acid hydrolysis in native urine samples, enzymatic hydrolysis produces markedly lower hydrolysis yields, which are between 41% and 74% in relation to the yield from acid hydrolysis. It is to be assumed that enzymatic hydrolysis cleaves the conjugates only incompletely. This incomplete cleavage cannot be quantified exactly, as the glucuronic acid conjugates are not commercially available. Therefore, in this context, acid hydrolysis – in spite of relatively high losses due to sample preparation – is recommended for conjugate cleavage.

10 Discussion of the method

The method described here represents a suitable and validated procedure for the determination of 2-phenyl-2-propanol in urine, by which both the conjugated and the non-conjugated fraction can be determined. With the present method a reliable determination of occupational cumene exposure is possible, as the 2-phenyl-2-propanol concentration in the urine behaves proportionally to the external exposure [13].

Calibration should be carried out in pooled urine, as matrix effects cannot be excluded with certainty in the case of calibration standards spiked in water. When using the urine of occupationally non-exposed individuals in preparing the pooled urine, no background signals above the detection limits are observed. The method is uncomplicated and using GC-FID it makes use of a widely distributed analytical technique. The obtained precision data are to be described as good. However, the losses due to sample preparation using acid hydrolysis must be examined by the user of the method. The losses due to sample preparation should be lower when using an isotope-labelled standard; this would mean changing to GC-MS as detection device.

Instruments used:

- Gas chromatograph Varian CP3800 with flame ionisations detector and autosampler Varian 8200; chromatography software: Varian Star 1

11 References

1 K. Weissermel and H.-J. Arpe (eds.): Industrielle organische Chemie. VCH Weinheim, 4th edition, 374 (1994).

2 F. Ullmann (ed.): Enzyklopädie der Technischen Chemie Band 10. VCH Weinheim, 4th edition (1990).

3 J. Falbe and M. Regitz (eds.): Römpp Lexikon Chemie. Thieme Verlag Stuttgart, 10th edition (1999).

4 Greim H. (ed.): Cumene. MAK Value Documentations. Volume 13, Wiley-VCH, Weinheim (1999).

5 W. Senczuk and B. Litewka: Absorption of cumene through the respiratory tract and excretion of dimethylphenylcarbinol in urine. Br. J. Ind. Med. 33, 100–105 (1976).

6 A.M. El Masry, J.N. Smith, R.T. Williams: Studies in detoxication. 69. The metabolism of alkylbenzenes: n-propylbenzene and n-butylbenzene with further observations on ethylbenzene. Biochem. J. 64, 50–56 (1956).

7 A. Sato and T. Nakajima: Partition coefficients of some aromatic hydrocarbons and ketones in water, blood and oil. Br. J. Ind. Med. 36, 231–234 (1979).

8 H.W. Gerarde: Toxicological studies on hydrocarbons. III. The biochemorphology of the phenylalkanes and phenylalkenes. AMA Arch. Ind. Health. 19, 403–418 (1959).

9 D. Robinson, J.N. Smith, R.T. Williams: Proceedings of the Biochemical Society: The metabolism of iso-propylbenzene (cumene). Biochem. J. 56, XI (1953).

10 K. Pyykkö, S. Paavilainen, T. Metsä-Ketelä, K. Laustiola: The increasing and decreasing effects of aromatic hydrocarbon solvents in pulmonary and hepatic cytochrome P-450 in rat. Pharmacol. Toxicol. 60, 288–293 (1987).

11 O.M. Bakke, R.R. Scheline: Hydroxylation of aromatic hydrocarbons in the rat. Toxicol. Appl. Pharmacol. 16, 691–700 (1970).

12 U. Knecht and A. Ulshöfer: Bio-Monitoring nach iso-Propylbenzoleinwirkung. 36. Jahrestagung der DGAUM. Wiesbaden. Rindt-Druck, Fulda; 225–227 (1996).

13 H. Greim and G. Lehnert (eds.): iso-Propylbenzol. Biologische Arbeitsstoff-Toleranz-Werte (BAT-Werte) und Expositionsäquivalente für krebserzeugende Arbeitsstoffe (EKA), 10th issue, Wiley-VCH, Weinheim (2001).

14 Bundesärztekammer: Richtlinie der Bundesärztekammer zur Qualitätssicherung quantitativer laboratoriumsmedizinischer Untersuchungen. Dt. Ärztebl. 105, A341–A355 (2008).

15 J. Angerer and G. Lehnert: Anforderungen an arbeitsmedizinisch-toxikologische Analysen – Stand der Technik. Dt. Ärztebl. 37, C1753–C1760 (1997).

16 J. Angerer, Th. Göen, G. Lehnert: Mindestanforderungen an die Qualität von umweltmedizinisch-toxikologischen Analysen. Umweltmed. Forsch. Prax. 3, 307–312 (1998).

Author: *U. Knecht*
Examiners: *Th. Göen, M. Müller*

12 Appendix

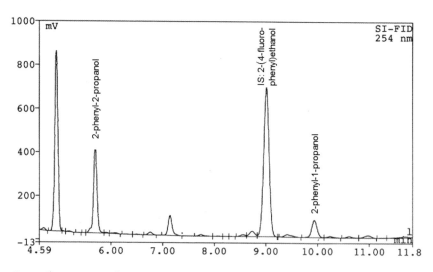

Fig. 2 Chromatogram of a processed urine sample of a volunteer exposed to cumene.

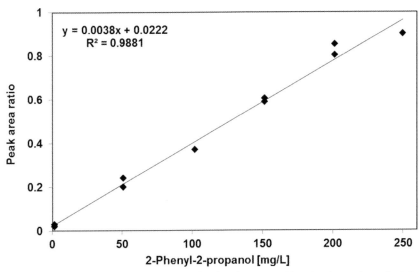

Fig. 3 Calibration graph for determination of 2-phenyl-2-propanol in urine. Internal standard: 2-(4-fluorophenyl)ethanol.

Pyrethrum and pyrethroid metabolites (after solid phase extraction) in urine

Matrix:	Urine
Hazardous substances:	Pyrethrum and pyrethroids
Analytical principle:	Capillary gas chromatography with mass selective detection (GC-MS)
Completed in:	November 2005

Overview of parameters that can be determined with this method and the corresponding chemical hazardous substances:

Substance	CAS	Parameter	CAS
Allethrin	584-79-2	trans-CDCA	72120-98-0
Cyfluthrin	68359-37-5	cis-DCCA	59042-49-8
		trans-DCCA	59042-50-1
		FPBA	77279-89-1
Cypermethrin	52315-07-8	cis-DCCA	59042-49-8
		trans-DCCA	59042-50-1
		3-PBA	3739-38-6
Deltamethrin	52918-63-5	cis-DBCA	63597-73-9
		3-PBA	3739-38-6
Permethrin	52645-53-1	cis-DCCA	59042-49-8
		trans-DCCA	59042-50-1
		3-PBA	3739-38-6
Phenothrin	26002-80-2	trans-CDCA	72120-98-0
Pyrethrum	8003-34-7	trans-CDCA	72120-98-0
Resmethrin	10453-86-8	trans-CDCA	72120-98-0
Tetramethrin	7696-12-0	trans-CDCA	72120-98-0

Summary

The present method is used to determine trans-chrysanthemum dicarboxylic acid (trans-CDCA) as metabolite of the pyrethroids allethrin, phenothrin, resmethrin, tetramethrin and of pyrethrum as well as cis- and trans-3-(2,2-dichlorovinyl)-2,2-dimethyl-cyclopropane carboxylic acid (cis- and trans-DCCA), cis-3-(2,2-dibromovinyl)-2,2-dimethyl-cyclopropane carboxylic acid (cis-DBCA), 4-fluoro-3-phenoxybenzoic acid (FPBA) and 3-phenoxybenzoic acid (3-PBA) as metabolites of the pyrethroids permethrin, cyfluthrin, cypermethrin and deltamethrin. Additionally, the presented method enables the determination of cis-chrysanthemum dicarboxylic acid (cis-CDCA).

With the gas chromatographic/mass spectrometric method presented here, the urinary concentration of seven pyrethroid and pyrethrum metabolites can be determined sensitively and reliably in one analytical procedure. Due to its sensitivity the procedure is suitable for detecting exposures relevant to both environmental medicine and occupational medicine.

The method comprises alkaline cleavage of the conjugates, the extraction of the carboxylic acids using solid phase extraction, as well as derivatisation by esterisation with hexafluoroisopropanol and diisopropyl carbodiimide. The analytical determination is carried out by gas chromatography/mass spectrometry after negative chemical ionisation. Quantitation is performed by use of two internal standards.

trans-Chrysanthemum dicarboxylic acid (trans-CDCA)

Within day precision:	Standard deviation (rel.)	s_w = 5.6% or 3.0%
	Prognostic range	u = 14.4% or 7.7%
	at a spiked concentration of 0.5 or 5.0 µg trans-CDCA per litre urine and where n = 6 determinations	
Day to day precision:	Standard deviation (rel.)	s_w = 3.7%
	Prognostic range	u = 9.5%
	at a spiked concentration of 0.5 or 5.0 µg trans-CDCA per litre urine and where n = 6 determinations	
Accuracy:	Recovery rate (rel.)	r = 93% or 98%
	at a nominal concentration of 0.5 or 5.0 µg trans-CDCA per litre urine and where n = 6 determinations	
Detection limit:	0.03 µg trans-CDCA per litre urine	
Quantitation limit:	0.09 µg trans-CDCA per litre urine	

cis-Chrysanthemum dicarboxylic acid (cis-CDCA)

Within day precision:	Standard deviation (rel.)	s_w = 6.7% or 5.4%
	Prognostic range	u = 17.2% or 13.9%
	at a spiked concentration of 0.5 or 5.0 µg cis-CDCA per litre urine and where n = 6 determinations	

Day to day precision: Standard deviation (rel.) s_w = 9.4% or 8.4%
Prognostic range u = 24.2% or 21.6%
at a spiked concentration of 0.5 or 5.0 µg cis-CDCA per litre urine and where n = 6 determinations

Accuracy: Recovery rate (rel.) r = 93% or 100%
at a nominal concentration of 0.5 or 5.0 µg cis-CDCA per litre urine and where n = 6 determinations

Detection limit: 0.03 µg cis-CDCA per litre urine
Quantitation limit: 0.09 µg cis-CDCA per litre urine

trans-3-(2,2-Dichlorovinyl)-2,2-dimethylcyclopropane carboxylic acid (trans-DCCA)

Within day precision: Standard deviation (rel.) s_w = 6.5% or 2.5%
Prognostic range u = 16.7% or 6.4%
at a spiked concentration of 0.5 or 5.0 µg trans-DCCA per litre urine and where n = 6 determinations

Day to day precision: Standard deviation (rel.) s_w = 7.4% or 1.6%
Prognostic range u = 19.0% or 4.1%
at a spiked concentration of 0.5 or 5.0 µg trans-DCCA per litre urine and where n = 6 determinations

Accuracy: Recovery rate (rel.) r = 91% or 105%
at a nominal concentration of 0.5 or 5.0 µg trans-DCCA per litre urine and where n = 6 determinations

Detection limit: 0.08 µg trans-DCCA per litre urine
Quantitation limit: 0.24 µg trans-DCCA per litre urine

cis-3-(2,2-Dichlorovinyl)-2,2-dimethylcyclopropane carboxylic acid (cis-DCCA)

Within day precision: Standard deviation (rel.) s_w = 5.4% or 1.5%
Prognostic range u = 13.9% or 3.9%
at a spiked concentration of 0.5 or 5.0 µg cis-DCCA per litre urine and where n = 6 determinations

Day to day precision: Standard deviation (rel.) s_w = 6.8% or 1.7%
Prognostic range u = 17.5% or 4.4%
at a spiked concentration of 0.5 or 5.0 µg cis-DCCA per litre urine and where n = 6 determinations

Accuracy: Recovery rate (rel.) r = 94%
at a nominal concentration of 0.5 or 5.0 µg cis-DCCA per litre urine and where n = 6 determinations

Detection limit: 0.08 µg cis-DCCA per litre urine
Quantitation limit: 0.24 µg cis-DCCA per litre urine

cis-3-(2,2-Dibromovinyl)-2,2-dimethylcyclopropane carboxylic acid (cis-DBCA)

Within day precision:	Standard deviation (rel.)	s_w = 3.1% or 1.6%
	Prognostic range	u = 8.0% or 4.1%
	at a spiked concentration of 0.5 or 5.0 µg cis-DBCA per litre urine and where n = 6 determinations	
Day to day precision:	Standard deviation (rel.)	s_w = 5.0% or 1.6%
	Prognostic range	u = 12.9% or 4.1%
	at a spiked concentration of 0.5 or 5.0 µg cis-DBCA per litre urine and where n = 6 determinations	
Accuracy:	Recovery rate (rel.)	r = 103% or 97%
	at a nominal concentration of 0.5 or 5.0 µg cis-DBCA per litre urine and where n = 6 determinations	
Detection limit:	0.06 µg cis-DBCA per litre urine	
Quantitation limit:	0.18 µg cis-DBCA per litre urine	

3-Phenoxybenzoic acid (3-PBA)

Within day precision:	Standard deviation (rel.)	s_w = 4.1% or 1.5%
	Prognostic range	u = 10.5% or 3.9%
	at a spiked concentration of 0.5 or 5.0 µg 3-PBA per litre urine and where n = 6 determinations	
Day to day precision:	Standard deviation (rel.)	s_w = 3.7% or 1.7%
	Prognostic range	u = 9.5% or 4.4%
	at a spiked concentration of 0.5 or 5.0 µg 3-PBA per litre urine and where n = 6 determinations	
Accuracy:	Recovery rate (rel.)	r = 100% or 99%
	at a nominal concentration of 0.5 or 5.0 µg 3-PBA per litre urine and where n = 6 determinations	
Detection limit:	0.15 µg 3-PBA per litre urine	
Quantitation limit:	0.45 µg 3-PBA per litre urine	

4-Fluoro-3-phenoxybenzoic acid (FPBA)

Within day precision:	Standard deviation (rel.)	s_w = 5.2% or 2.3%
	Prognostic range	u = 13.4% or 5.9%
	at a spiked concentration of 0.5 or 5.0 µg FPBA per litre urine and where n = 6 determinations	
Day to day precision:	Standard deviation (rel.)	s_w = 5.2% or 1.5%
	Prognostic range	u = 13.4% or 3.9%
	at a spiked concentration of 0.5 or 5.0 µg FPBA per litre urine and where n = 6 determinations	
Accuracy:	Recovery rate (rel.)	r = 92% or 98%
	at a nominal concentration of 0.5 or 5.0 µg FPBA per litre urine and where n = 6 determinations	
Detection limit:	0.04 µg FPBA per litre urine	
Quantitation limit:	0.12 µg FPBA per litre urine	

Pyrethroids and pyrethrum

A comprehensive presentation of general toxicological information on pyrethrum and the pyrethroids can be found in the method entitled "Pyrethrum and pyrethroid metabolites (after liquid-liquid extraction) in urine" (Authors: Leng, Gries) published in this book.

A summary of the toxicological properties of pyrethroids and pyrethrum can be found in the BAT Documentation entitled "Pyrethroids" [1] as well as the MAK Documentations on cyfluthrin and pyrethrum [2, 3]. The monograph on pyrethroids by the "Agency for Toxic Substances and Disease Registry" (USA) is additionally recommended [4].

Authors: *E. Berger-Preiß, S. Gerling*
Examiner: *D. Barr*

Pyrethrum and pyrethroid metabolites (after solid phase extraction) in urine

Matrix:	Urine
Hazardous substances:	Pyrethrum and pyrethroids
Analytical principle:	Capillary gas chromatography with mass selective detection (GC-MS)
Completed in:	November 2005

Contents

1	General principles
2	Equipment, chemicals and solutions
2.1	Equipment
2.2	Chemicals
2.3	Solutions
2.4	Internal standards
2.5	Calibration standards
3	Specimen collection and sample preparation
3.1	Specimen collection
3.2	Sample preparation
4	Operational parameters
4.1	Operational parameters for gas chromatography
4.2	Operational parameters for mass spectrometry
5	Analytical determination
6	Calibration
7	Calculation of the analytical result
8	Standardisation and quality control
9	Evaluation of the method
9.1	Precision

9.2 Accuracy
9.3 Detection limits and quantitation limits
9.4 Sources of error
10 Discussion of the method
11 References
12 Appendix

1 General principles

The urine samples are subjected to alkaline hydrolysis after addition of the internal standards. Separation of the analytes from matrix components is carried out by solid phase extraction. The carboxylic acids are then derivatised with hexafluoroisopropanol and diisopropyl carbodiimide and determined analytically after capillary gas chromatographic separation with a mass selective detector following chemical ionisation.

2 Equipment, chemicals and solutions

2.1 Equipment

- Gas chromatograph with split/splitless injector, mass selective detector, autosampler and data processing system
- Capillary gas chromatographic column: length: 60 m; inner diameter: 0.25 mm; stationary phase: (5%-phenyl)-methylpolysiloxane; film thickness: 0.25 µm (e.g. HP5 MS, Agilent No. 19091S-436E)
- Columns for solid phase extraction: Waters Oasis MAX columns 6 mL/150 mg (Order No. 186000369)
- Vacuum column holder for solid phase extraction with evaporation unit using nitrogen (e.g. Supelco Visiprep)
- 12 mL, 20 mL and 25 mL centrifuge glass tubes (Duranglas) with teflon-coated screw caps (e.g. Schütt)
- Polypropylene containers for urine collection (e.g. Kautex)
- Analytical balance (e.g. Sartorius)
- Laboratory shaker (e.g. IKA Vibrax VXR)
- Various glass beakers
- Various volumetric flasks
- Microlitre pipettes (e.g. Eppendorf)
- 1.8 mL crimp cap vials for GC (e.g. Agilent)
- Drying oven

Note:

In order to avoid analyte losses from adsorption on glass walls, silanised glass equipment is to be used. Silanisation is carried out with Sylon CT. For this purpose, the glass equipment is soaked with the silylation reagent and incubated for about 30 min. Subsequently, the glass equipment is consecutively rinsed with hexane, toluene and methanol, and then dried.

2.2 Chemicals

- E-cis-/trans-Chrysanthemum dicarboxylic acid (e.g. Synchem No. a047)[1)]
- cis-/trans-3-(2,2-Dichlorovinyl)-2,2-dimethylcyclopropane carboxylic acid (e.g. LGC standard No. CIL-ULM-7303)[1)]
- cis-3-(2,2-Dibromovinyl)-2,2-dimethylcyclopropane carboxylic acid (e.g. Ehrenstorfer No. 12244000)
- 4-Fluoro-3-phenoxybenzoic acid (e.g. Promochem No. CIL-ULM-7391)
- 3-Phenoxybenzoic acid (e.g. Sigma-Aldrich No. 77708)
- 2-Phenoxybenzoic acid (e.g. Sigma-Aldrich No. 153176)
- 1,1,1,3,3,3-Hexafluoroisopropanol 99.8% (e.g. Sigma-Aldrich No. 52517)
- N,N'-Diisopropyl carbodiimide 99% (e.g. Sigma-Aldrich No. D125407)
- Purified water (e.g. Milli-Q-water)
- Acetonitrile anhydrous (e.g. Merck No. 100004)
- p-Phenylene diacetic acid ≥ 97% (e.g. Sigma-Aldrich No. 78470)
- Ethyl acetate ≥ 99.8% (e.g. Sigma-Aldrich No. 439169)
- Methanol p.a. (e.g. Merck No. 106009)
- n-Hexane (e.g. Sigma-Aldrich No. 296090)
- Hydrochloric acid 0.1 N (e.g. Merck No. 109060)
- Hydrochloric acid, concentrated, 37% (e.g. Merck No. 100317)
- Potassium carbonate p.a. (e.g. Merck No. 104928)
- Potassium hydroxide p.a. (e.g. Merck No. 105033)
- Toluene (e.g. Merck No. 108325)
- Ammonium hydroxide solution 25% p.a. (e.g. Merck No. 105432)
- Sylon-CT, 5% dimethyldichlorosilane in toluene (e.g. Sigma-Aldrich No. 33065-U)
- Nitrogen 5.0 (e.g. Linde)
- Helium 5.0 (e.g. Linde)
- Methane 5.5 (e.g. Linde)

---

1) Classification and quantitation of the isomers is performed after HPLC separation via ^{1}H-NMR [8].

2.3 Solutions

- 10 M Potassium hydroxide
  112.2 g potassium hydroxide pellets are weighed exactly into a 250 mL glass beaker and dissolved in about 100 mL highly purified water. The solution is transferred to a 200 mL volumetric flask. The flask is made up to the mark with highly purified water.
  At room temperature the solution can be kept for at least 6 months.
- 2% Ammonium hydroxide solution
  80 mL of the 25% ammonium hydroxide solution are placed into a 1000 mL volumetric flask. The flask is made up to the mark with highly purified water.
  At room temperature the solution can be kept for at least 6 months.
- 5% Methanol in ethyl acetate
  25 mL methanol are pipetted into a 500 mL volumetric flask, which is made up to the mark with ethyl acetate.
  At room temperature the solution can be kept for at least 1 week.
- 5% Potassium carbonate solution
  25 g Potassium carbonate are weighed exactly into a 500 mL volumetric flask, dissolved in highly purified water and the flask is made up to the mark with water.
  At room temperature the solution can be kept for at least 6 months.

2.4 Internal standards

- Stock solution of the internal standards (1 g/L)
  10 mg 2-phenoxybenzoic acid (2-PBA) and 10 mg p-phenylene diacetic acid (PDAA) are weighed exactly into a 10 mL volumetric flask, which is made up to the mark with acetonitrile.
- Working solution of the internal standards (1 mg/L)
  0.01 mL of the stock solution of 2-PBA and PDAA are pipetted into a 10 mL volumetric flask, which is made up to the mark with highly purified water.

Both solutions can be kept in the refrigerator (4°C) for at least 6 months.

2.5 Calibration standards

- Stock solutions (1 g/L)
  10 mg each of trans- and cis-CDCA, trans- and cis-DCCA, cis-DBCA, FPBA and 3-PBA are weighed exactly into a 10 mL volumetric flask each, which is made up to the mark with acetonitrile.

- Intermediate dilution (10 mg/L)
  0.1 mL each of the different stock solutions are pipetted into a 10 mL volumetric flask, which is made up to the mark with acetonitrile.
- Working solution A (1 mg/L)
  1 mL of the intermediate dilution is pipetted into a 10 mL volumetric flask, which is made up to the mark with highly purified water.
- Working solution B (100 µg/L)
  0.1 mL of the intermediate dilution is pipetted into a 10 mL volumetric flask, which is made up to the mark with highly purified water.

The solutions can be kept in the refrigerator (4°C) for at least 6 months.

The calibration solutions are prepared in urine. For this purpose the working solutions A and B are diluted with pooled urine in a 20 mL screw cap glass tube according to the pipetting scheme given in Table 1. The calibration solutions are then prepared and analysed analogously to the urine samples as described in Section 3 and Section 4.

Table 1 Pipetting scheme for the preparation of calibration solutions.

Calibration solution	Volume of working solution [µL] A	Volume of working solution [µL] B	Addition of pooled urine [µL]	Concentration of calibration solution [µg/L]
K0	–	–	5000	0
K1	–	10	4990	0.2
K2	–	25	4975	0.5
K3	–	50	4950	1.0
K4	25	–	4975	5.0
K5	50	–	4950	10.0

The calibration solutions are freshly prepared for each analytical series. At –20°C the solutions can be stored for up to a year.

3 Specimen collection and sample preparation

3.1 Specimen collection

The urine is collected in sealable plastic containers and should be stored at –20°C before analysis. Under these storage conditions, the urine is stable for at least one year. When stored in the refrigerator (4°C) the urine can be kept for at least five weeks, if it is preserved with 1% chloroform.

3.2 Sample preparation

5 mL urine are pipetted into a 25 mL screw cap glass tube and diluted with 5 mL water. After addition of 20 µL of the spiking solution of the internal standards (corresponding to a concentration of 4 µg/L each of 2-PBA and PDAA in urine) 1 mL 10 M potassium hydroxide is added, the tube is sealed and heated for 15 min to 70°C in the drying oven (hydrolysis step). After equilibration of the solution to room temperature, 1 mL concentrated hydrochloric acid and 10 mL water are added.

For solid phase extraction each cartridge is initially conditioned consecutively with 6 mL each of ethyl acetate, methanol, water and 0.1 N hydrochloric acid. Then the entire sample solution is transferred in several portions with a pasteur pipette to the conditioned cartridge. Subsequently, the cartridge is washed with 6 mL 0.1 N hydrochloric acid and then dried under nitrogen. Then, the cartridge is washed with 6 mL 2% ammonium hydroxide solution and consecutively rinsed with 6 mL methanol and 6 mL ethyl acetate. The cartridge is dried with nitrogen, then washed with 6 mL 0.1 N hydrochloric acid for acidification and dried again with nitrogen. The analytes are finally eluted in a new 12 mL glass tube (silanised) using 6 mL 5% methanol in ethyl acetate.

The eluate is evaporated to dryness with nitrogen and the residue is dissolved in 5 mL n-hexane. Subsequently, 50 µL hexafluoroisopropanol and 75 µL diisopropyl carbodiimide are added for derivatisation. The solution is mixed for 30 min on a laboratory shaker. Then, a surplus of the derivatisation reagent is removed by adding 5 mL 5% potassium carbonate solution (mix for 5 min on a laboratory shaker). An aliquot of 1 mL of the organic phase is transferred to a GC vial and 1 µL of the sample solution is injected splitless into the GC-MS system. To increase sensitivity, the sample solution can be further concentrated by reducing the hexane phase to half by evaporation in a stream of nitrogen. The prepared samples are stable at 4°C for at least 14 days. Longer storage times were not tested.

4 Operational parameters

4.1 Operational parameters for gas chromatography

Capillary column:
- Material: Fused Silica
- Stationary phase: HP 5 MS
- Length: 60 m
- Inner diameter: 0.25 mm
- Film thickness: 0.25 µm

Temperatures:	Column:	Initial temperature 50°C, then increase at a rate of 10°C/min to 130°C, increase at 5°C/min to 190°C, increase at 25°C/min to 280°C, 3 min at final temperature
	Injector:	250°C
	Transfer line:	280°C
Carrier gas:	Helium 5.0, constant flow of 1.4 mL/min	
Split:	splitless	
Injection volume:	1 µL	

4.2 Operational parameters for mass spectrometry

Ionisation type:	Negative chemical ionisation (NCI)
Reactant gas:	Methane, 40%
Ionisation energy:	96 eV
Source temperature:	150°C
Multiplier:	2518 V

All other parameters must be optimised in accordance with the manufacturer's instructions.

5 Analytical determination

For analytical determination of the urine samples prepared according to Section 3.2, 1 µL of the n-hexane extract is injected into the gas chromatograph. The temporal profiles of the ion traces listed in Table 2 are recorded in the SIM mode.

The retention times given are intended to be a rough guide only. Users of the method must ensure proper separation performance of the capillary column used and the resulting retention behaviour of the analytes.

The CI mass spectra of the analytes are illustrated in the Appendix (see Figure 3). A chromatogram of a pooled urine sample spiked with standard solutions is also shown (see Figure 2).

Table 2 Retention times and recorded masses of the monoesterified (*) or diesterified (**) analytes.

Analyte	Retention time [min]	Recorded mass [m/z] Quantifier	Qualifier
trans-CDCA**	11.56	498	330
cis-CDCA**	12.16	498	330
cis-DCCA*	11.88	322	324
trans-DCCA*	11.97	322	324
cis-DBCA*	15.05	366	81
FPBA*	19.47	231	382
3-PBA*	20.19	364	213
PDAA** (IS)	14.80	298	326
2-PBA* (IS)	19.85	364	213

6 Calibration

The calibration standards are prepared and processed according to Section 2.5 and Section 3 and are analysed as described in Section 4 and Section 5.

The calibration graphs are obtained by plotting the quotients of the peak areas of the analytes and the corresponding internal standards against the spiked concentration used. For evaluation of the diesterified metabolites trans- and cis-CDCA, PDAA is used as internal standard, as PDAA is also twofold esterised. 2-PBA (monoesterified) is used as internal standard for the other analytes.

The linearity of the method was tested and confirmed in the range from 0.2 µg/L to 10 µg/L. Any blank values which may still be present are accounted for by subtraction.

The calibration graphs of the individual metabolites are given in Figure 1 of the Appendix.

7 Calculation of the analytical result

To determine the analyte concentration in a sample, the peak area of the analyte is divided by the peak area of the respective internal standard. The quotient thus obtained is inserted into the corresponding calibration graph. By dividing the quotient by the slope of the calibration graph, the analyte concentration level of the sample is obtained in µg/L urine. If the determined analyte level of the sample is outside the calibration range, the urine sample is diluted and prepared again.

8 Standardisation and quality control

Quality control of the analytical results is carried out as stipulated in the guidelines of the *Bundesärztekammer* (German Medical Association) [5–7].

To check precision urine spiked with a known and constant analyte concentration is analysed as control sample within each analytical series. As a material for quality control is not commercially available, it must be prepared in the laboratory by adding a defined quantity of the metabolites to be investigated to pooled urine. The thus obtained reference urine is divided into aliquots and kept at –20°C. The concentration of this control material should lie within the relevant concentration range. The nominal value and the tolerance ranges of this quality control material are determined in a pre-analytical period [7]. The measured values of the control samples analysed within each analytical series should lie within the tolerance ranges obtained.

9 Evaluation of the method

9.1 Precision

To determine within day precision pooled urine is spiked with defined quantities of the metabolites to obtain concentration levels of 0.5 µg/L and 5.0 µg/L. These

Table 3 Within day precision for the determination of trans- and cis-CDCA, trans- and cis-DCCA, cis-DBCA, 3-PBA and FPBA in urine based on two spiked control urine samples (n = 6).

Analyte		Spiked concentration	
		0.5 µg/L	5.0 µg/L
trans-CDCA	standard deviation (rel.) s_w [%]	5.6	3.0
	prognostic range u [%]	14.4	7.7
cis-CDCA	standard deviation (rel.) s_w [%]	6.7	5.4
	prognostic range u [%]	17.2	13.9
cis-DCCA	standard deviation (rel.) s_w [%]	5.4	1.5
	prognostic range u [%]	13.9	3.9
trans-DCCA	standard deviation (rel.) s_w [%]	6.5	2.5
	prognostic range u [%]	16.7	6.4
cis-DBCA	standard deviation (rel.) s_w [%]	3.1	1.6
	prognostic range u [%]	8.0	4.1
3-PBA	standard deviation (rel.) s_w [%]	4.1	1.5
	prognostic range u [%]	10.5	3.9
FPBA	standard deviation (rel.) s_w [%]	5.2	2.3
	prognostic range u [%]	13.4	5.9

urine samples were prepared and analysed six times in a row. The obtained within day precisions are documented in Table 3.

Furthermore, the day to day precision was determined. For this purpose the two spiked urine samples were also used. These were prepared and analysed on six different days. The precisions obtained in this manner are shown in Table 4.

Table 4 Day to day precision for the determination of trans- and cis-CDCA, trans- and cis-DCCA, cis-DBCA, 3-PBA and FPBA in urine based on two spiked control urine samples (n = 6).

Analyte		Spiked concentration	
		0.5 µg/L	5.0 µg/L
trans-CDCA	standard deviation (rel.) s_w [%]	3.7	3.7
	prognostic range u [%]	9.5	9.5
cis-CDCA	standard deviation (rel.) s_w [%]	9.4	8.4
	prognostic range u [%]	24.2	21.6
cis-DCCA	standard deviation (rel.) s_w [%]	6.8	1.7
	prognostic range u [%]	17.5	4.4
trans-DCCA	standard deviation (rel.) s_w [%]	7.4	1.6
	prognostic range u [%]	19.0	4.1
cis-DBCA	standard deviation (rel.) s_w [%]	5.0	1.6
	prognostic range u [%]	12.9	4.1
3-PBA	standard deviation (rel.) s_w [%]	3.7	1.7
	prognostic range u [%]	9.5	4.4
FPBA	standard deviation (rel.) s_w [%]	5.2	1.5
	prognostic range u [%]	13.4	3.9

9.2 Accuracy

To determine the accuracy of the method recovery experiments were carried out. For this purpose pooled urine was spiked with defined concentrations (0.5 µg/L and 5.0 µg/L) of the metabolites to be investigated. The samples were processed and analysed six times as described in Section 3.2. The relative recovery rates obtained are summarised in Table 5.

9.3 Detection limits and quantitation limits

The detection limit was estimated on the basis of a signal to noise ratio of 3:1. The quantitation limit was calculated on the basis of a signal to noise ratio of 9:1. The values thus obtained are summarised in Table 6.

Table 5 Relative recovery rates of the analytes cis- and trans-CDCA, cis- and trans-DCCA, cis-DBCA, 3-PBA and FPBA in two spiked urine samples (n = 6).

Analyte		Spiked concentration	
		0.5 µg/L	5.0 µg/L
trans-CDCA	relative recovery r [%]	93	98
	range [%]	82–104	84–106
cis-CDCA	relative recovery r [%]	93	100
	range [%]	87–96	94–103
cis-DCCA	relative recovery r [%]	94	94
	range [%]	84–102	90–95
trans-DCCA	relative recovery r [%]	91	105
	range [%]	90–95	103–108
cis-DBCA	relative recovery r [%]	103	97
	range [%]	94–106	94–99
3-PBA	relative recovery r [%]	100	99
	range [%]	93–106	97–101
FPBA	relative recovery r [%]	92	98
	range [%]	83–99	96–100

Table 6 Detection and quantitation limits of the analytes.

Analyte	Detection limit [µg/L]	Quantitation limit [µg/L]
trans-CDCA	0.03	0.09
cis-CDCA	0.03	0.09
trans-DCCA	0.08	0.24
cis-DCCA	0.08	0.24
cis-DBCA	0.06	0.18
3-PBA	0.15	0.45
FPBA	0.04	0.12

9.4 Sources of error

To avoid analyte losses by adsorption on glass walls, the use of silanised glass equipment is advantageous. In the present method silanisation is carried out with the silylation reagent Sylon CT (see Section 2.1).

The presented solid phase extraction procedure involves relatively complex washing steps. Pre-cleaning of the cartridges with ethyl acetate and methanol is recommended to avoid possible subsequent contaminations from cartridge impurities. Washing with water and 0.1 N hydrochloric acid serves to condition the cartridge before sample introduction, as solvent residues in the cartridge can negatively affect the retention behaviour of the analytes.

After transfer of the sample solution to the cartridge, the salts contained in the urine are eluted in a washing step using aqueous hydrochloric acid. Washing with methanol and ethyl acetate after alkaline treatment serves to remove further urine matrix components. This makes it almost impossible to effectively reduce the number of washing steps. All washing steps serve to remove urine matrix components which otherwise would be eluted with the analytes.

In order to recognize possible interferences from the urine matrix, especially when analysing the samples using GC-NCI/MS, care must be taken that the response ratio between the quantifier and qualifier mass traces agrees with that of a standard having a similar concentration (±30% tolerance).

10 Discussion of the method

With the described procedure both the metabolites formed from natural pyrethrum and those from the synthetic pyrethroids can be analysed within one analytical series.

Derivatisation of the analytes in urine takes place almost completely when using the described method. This was checked both by standard addition and by HPLC as well as by spiking a SPE extract that was free of blind values. As derivatisation takes place in a non-polar solvent, a relatively large volume of derivatisation solution is necessary to dissolve the residue obtained after SPE. The use of a smaller solvent volume is not recommended. Sensitivity can be increased if necessary by reducing the sample volume after derivatisation.

In addition the suitability of 3-methylene-1,2-cyclopropane dicarboxylic acid for use as internal standard (IS) for cis- and trans-CDCA was tested. However, due to the high losses during solid phase extraction and the relatively high volatility, another IS had to be used. PDAA was found to be suitable and is also commercially available. However its sensitivity is lower than that of cis- and trans-CDCA, as no molecular ion occurs in the mass spectrum (see Figure 3 in the Appendix). For this reason it is recommended to use a higher concentration of the internal standard PDAA, especially when analysing real samples.

Care must be taken that recovery should be checked again with spiked pooled urine when using a new batch of SPE cartridges.

After solid phase extraction the eluate is evaporated to dryness. In order to remove any remaining traces of water, a small amount of acetonitrile can be added to the eluate before evaporation. As acetonitrile forms an azeotrope with water in a ratio of 7:1, a faster evaporation of water is achieved.

Altogether, the present method is characterised by good precision, accuracy and sensitivity. It is therefore ideally suitable for application in the fields of occupational and environmental medicine.

Further investigations with the present method are described in the publication of Elflein et al. (2003) [8].

Instruments used:

- Gas chromatograph (HP 6890 Plus, Agilent, USA) with split/splitless injector and autosampler (MPS2, Gerstel, Germany) with a mass spectrometer (HP 5973, Agilent, USA). As software, HP Chemstation was used for measurement and evaluation.

11 References

1. H. Greim and H. Drexler (eds.): Pyrethroide. Biologische Arbeitsstoff-Toleranz-Werte (BAT-Werte) und Expositionsäquivalente für krebserzeugende Arbeitsstoffe (EKA) und Biologische Leitwerte (BLW). 15th issue, Wiley-VCH-Verlag, Weinheim (2008).
2. H. Greim (ed.): Pyrethrum. Gesundheitsschädliche Arbeitsstoffe, Toxikologisch-arbeitsmedizinische Begründung von MAK-Werten. 45th issue, Wiley-VCH Verlag, Weinheim (2008).
3. H. Greim (ed.): Cyfluthrin. The MAK Collection Part I: MAK Value Documentations. Volume 23, Wiley-VCH Verlag, Weinheim (2007).
4. Agency for Toxic Substances and Disease Registry: Toxicological profile for pyrethrins and pyrethroids. US Departement of Health and Human Services, Atlanta (2003).
5. Bundesärztekammer: Qualitätssicherung der quantitativen Bestimmungen im Laboratorium. Neue Richtlinien der Bundesärztekammer. Dt. Ärztebl. 85, A699–A712 (1988).
6. Bundesärztekammer: Ergänzung der "Richtlinien der Bundesärztekammer zur Qualitätssicherung in medizinischen Laboratorien". Dt. Ärztebl. 91, C159–C161 (1994).
7. J. Angerer, T. Göen, G. Lehnert: Mindestanforderungen an die Qualität von umweltmedizinisch-toxikologischen Analysen. Umweltmed. Forsch. Prax. 3, 307–312 (1998).
8. L. Elflein, E. Berger-Preiß, A. Preiß, M. Elend, K. Levsen, G. Wünsch: Human biomonitoring of pyrethrum and pyrethroid insecticides used indoors: Determination of the metabolites E-cis/trans-chrysanthemumdicarboxylic acid in human urine by gas chromatography-mass spectrometry with negative chemical ionization. J. Chromatogr. B 795, 195–207 (2003).

Authors: *E. Berger-Preiß, S. Gerling*
Examiner: *D. Barr*

12 Appendix

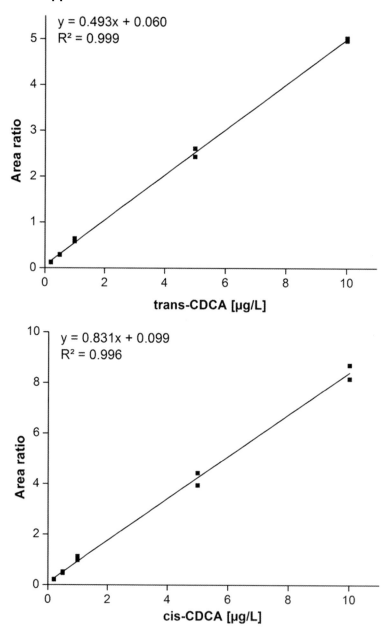

Fig. 1 a) Calibration graphs of the analysed metabolites in urine in a concentration range from 0.2 to 10 µg/L: trans-/cis-CDCA.

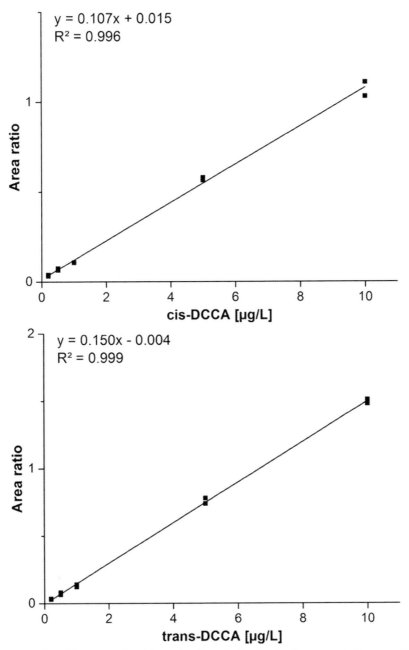

Fig. 1 b) Calibration graphs of the analysed metabolites in urine in a concentration range from 0.2 to 10 µg/L: cis-/trans-DCCA.

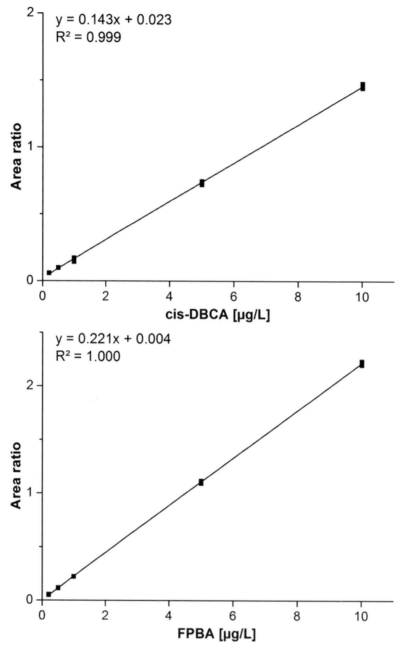

Fig. 1 c) Calibration graphs of the analysed metabolites in urine in a concentration range from 0.2 to 10 μg/L: cis-DBCA and FPBA.

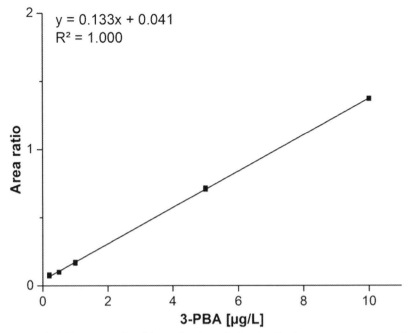

Fig. 1 d) Calibration graphs of the analysed metabolites in urine in a concentration range from 0.2 to 10 µg/L: 3-PBA.

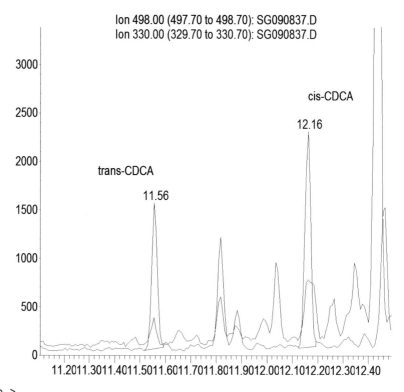

Fig. 2 Chromatograms of a urine sample spiked with 1 μg/L of the analysed metabolites (a–f). a) Hexafluorodiisopropyl ester of cis- and trans-CDCA.

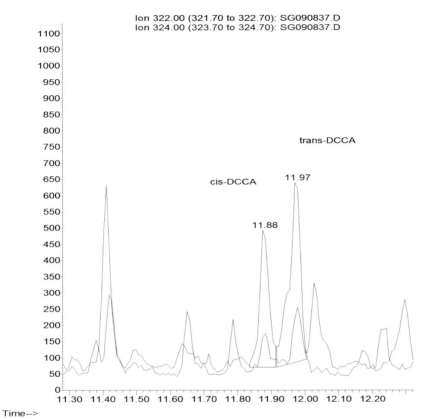

Fig. 2 b) Hexafluoroisopropyl ester of cis- and trans-DCCA.

204 | Analytical Methods

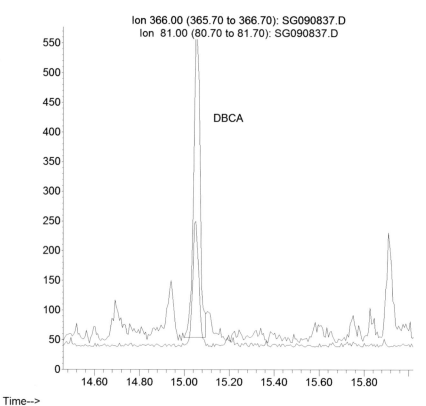

Fig. 2 c) cis-DBCA-hexafluoroisopropyl ester.

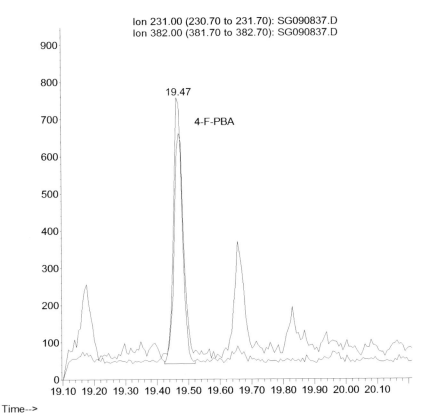

Fig. 2 d) FPBA-hexafluoroisopropyl ester.

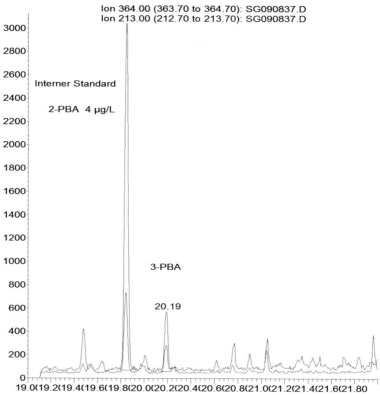

Fig. 2 e) Hexafluoroisopropyl ester of 2-PBA (IS) and 3-PBA.

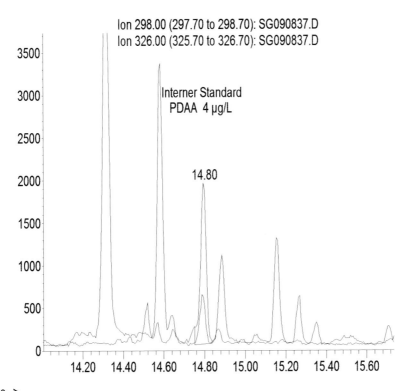

Fig. 2 f) Hexafluorodiisopropyl ester of PDAA (IS).

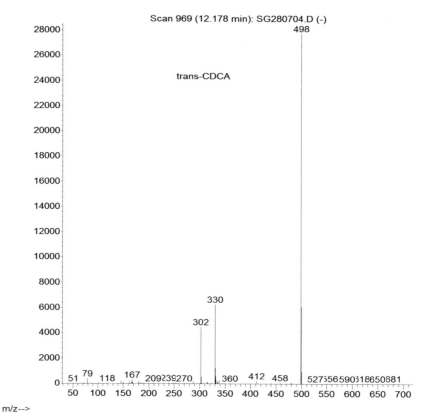

Fig. 3 GC-NCI-MS spectra of the analytes and of the internal standards (a–g). a) Hexafluorodiisopropyl ester of trans-CDCA.

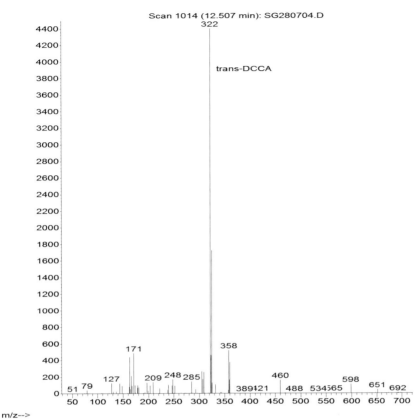

Fig. 3 b) Hexafluoroisopropyl ester of trans-DCCA.

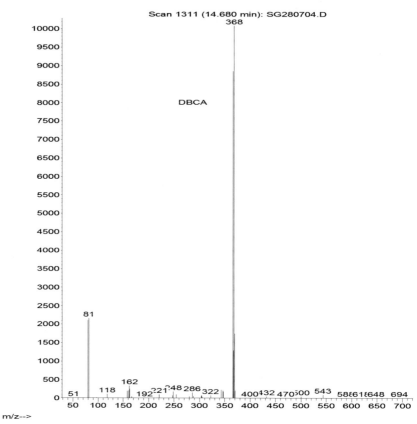

Fig. 3 c) Hexafluoroisopropyl ester of cis-DBCA.

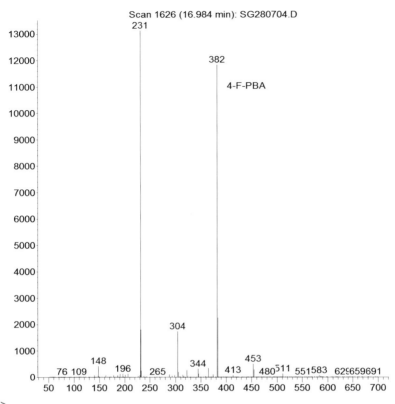

Fig. 3 d) Hexafluoroisopropyl ester of FPBA.

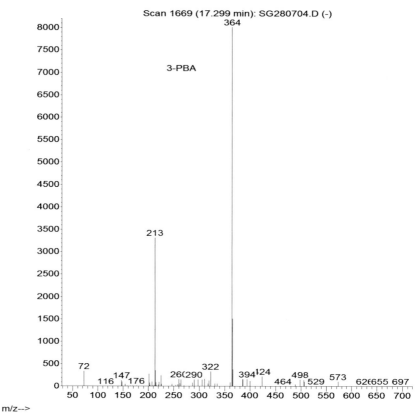

Fig. 3 e) Hexafluoroisopropyl ester of 3-PBA.

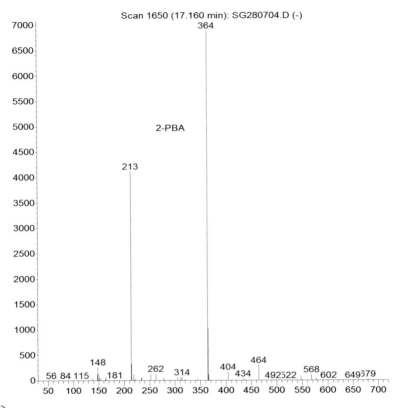

Fig. 3 f) Hexafluoroisopropyl ester of the internal standard 2-PBA.

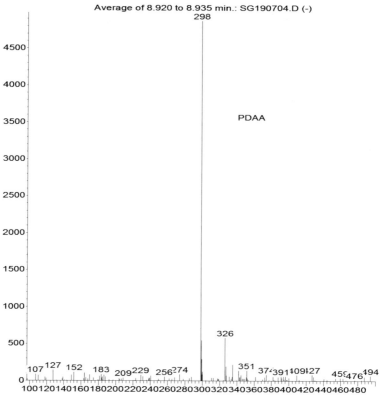

Fig. 3 g) Hexafluorodiisopropyl ester of the internal standard PDAA.

Pyrethrum and pyrethroid metabolites (after liquid-liquid extraction) in urine

Matrix:	Urine
Hazardous substances:	Pyrethrum and pyrethroids
Analytical principle:	Capillary gas chromatography with mass selective detection (GC-MS)
Completed in:	November 2005

Overview of parameters that can be determined with this method and the corresponding hazardous substances:

Substance	CAS	Parameter	CAS
Allethrin	584-79-2	trans-CDCA	72120-98-0
Cyfluthrin	68359-37-5	cis-DCCA	59042-49-8
		trans-DCCA	59042-50-1
		FPBA	77279-89-1
Cypermethrin	52315-07-8	cis-DCCA	59042-49-8
		trans-DCCA	59042-50-1
		3-PBA	3739-38-6
Deltamethrin	52918-63-5	cis-DBCA	63597-73-9
		3-PBA	3739-38-6
Permethrin	52645-53-1	cis-DCCA	59042-49-8
		trans-DCCA	59042-50-1
		3-PBA	3739-38-6
Phenothrin	26002-80-2	trans-CDCA	72120-98-0
Pyrethrum	8003-34-7	trans-CDCA	72120-98-0
Resmethrin	10453-86-8	trans-CDCA	72120-98-0
Tetramethrin	7696-12-0	trans-CDCA	72120-98-0

Summary

The method described here permits a simultaneous, sensitive and reliable determination of the concentration levels of six pyrethroid and pyrethrum metabolites in urine. On account of its sensitivity, the procedure is suitable for biomonitoring studies in both, environmental medicine and occupational medicine.

The present method is used to determine trans-chrysanthemum dicarboxylic acid (trans-CDCA) as metabolite of the pyrethroids allethrin, phenothrin, resmethrin, tetramethrin and pyrethrum as well as cis- and trans-3-(2,2-dichlorovinyl)-2,2-dimethylcyclopropane carboxylic acid (cis- and trans-DCCA), cis-3-(2,2-dibromo-vinyl)-2,2-dimethylcyclopropane carboxylic acid (cis-DBCA), 4-fluoro-3-phenoxy-benzoic acid (FPBA) and 3-phenoxybenzoic acid (3-PBA) as metabolites of the pyrethroids permethrin, cyfluthrin, cypermethrin and deltamethrin.

For determination, the conjugates of the metabolites in the urine are subjected to acid hydrolysis, separated using liquid/liquid extraction and subsequently derivatised with hexafluoroisopropanol and diisopropyl carbodiimide. After extraction of the derivates with isooctane, analytical determination is carried out by gas chromatographic separation with subsequent mass spectrometric detection after electron impact ionisation (EI mode) or negative chemical ionisation (NCI mode).

The following validation data relate to the method after analysis using high-resolution sector field mass spectrometry in the NCI mode. The characteristics of the method variant using the EI mode for ionisation and the reliability data of the alternative method (quadrupol mass spectrometry with NCI) are given in Section 9.

trans-Chrysanthemum dicarboxylic acid (trans-CDCA)

Within day precision: Standard deviation (rel.) s_w = 3.8%, 4.1% or 5.1%
Prognostic range u = 9.8%, 10.5% or 13.1%
at a spiked concentration of 0.2, 1.0 or 10.0 µg trans-CDCA per litre urine and where n = 6 determinations

Day to day precision: Standard deviation (rel.) s_w = 8.1%, 6.5% or 13.1%
Prognostic range u = 20.8%, 16.7% or 33.7%
at a spiked concentration of 0.2, 1.0 or 10.0 µg trans-CDCA per litre urine and where n = 6 determinations

Accuracy: Recovery rate (rel.) r = 90% or 107%
at a nominal concentration of 1.0 µg trans-CDCA per litre urine and where n = 6 determinations within day or from day to day

Detection limit: 0.05 µg trans-CDCA per litre urine
Quantitation limit: 0.15 µg trans-CDCA per litre urine

trans-3-(2,2-Dichlorovinyl)-2,2-dimethylcyclopropane carboxylic acid (trans-DCCA)

Within day precision:	Standard deviation (rel.)	s_w = 3.8%, 3.0% or 7.6%
	Prognostic range	u = 9.8%, 7.7% or 19.5%
	at a spiked concentration of 0.2, 1.0 or 10.0 µg trans-DCCA per litre urine and where n = 6 determinations	
Day to day precision:	Standard deviation (rel.)	s_w = 13.4%, 12.7% or 10.0%
	Prognostic range	u = 34.5%, 32.7% or 25.7%
	at a spiked concentration of 0.2, 1.0 or 10.0 µg trans-DCCA per litre urine and where n = 6 determinations	
Accuracy:	Recovery rate (rel.)	r = 106% or 102%
	at a nominal concentration of 1.0 µg trans-DCCA per litre urine and where n = 6 determinations within day or from day to day	
Detection limit:	0.02 µg trans-DCCA per litre urine	
Quantitation limit:	0.06 µg trans-DCCA per litre urine	

cis-3-(2,2-Dichlorovinyl)-2,2-dimethylcyclopropane carboxylic acid (cis-DCCA)

Within day precision:	Standard deviation (rel.)	s_w = 7.6%, 4.5% or 8.7%
	Prognostic range	u = 19.5%, 11.6% or 22.4%
	at a spiked concentration of 0.2, 1.0 or 10.0 µg cis-DCCA per litre urine and where n = 6 determinations	
Day to day precision:	Standard deviation (rel.)	s_w = 11.2%, 13.3% or 13.0%
	Prognostic range	u = 28.8%, 34.2% or 33.4%
	at a spiked concentration of 0.2, 1.0 or 10.0 µg cis-DCCA per litre urine and where n = 6 determinations	
Accuracy:	Recovery rate (rel.)	r = 94% or 90%
	at a nominal concentration of 1.0 µg cis-DCCA per litre urine and where n = 6 determinations within day or from day to day	
Detection limit:	0.02 µg cis-DCCA per litre urine	
Quantitation limit:	0.06 µg cis-DCCA per litre urine	

cis-3-(2,2-Dibromovinyl)-2,2-dimethylcyclopropane carboxylic acid (cis-DBCA)

Within day precision:	Standard deviation (rel.)	s_w = 6.0%, 4.6% or 9.3%
	Prognostic range	u = 15.4%, 11.8% or 23.9%
	at a spiked concentration of 0.2, 1.0 or 10.0 µg cis-DBCA per litre urine and where n = 6 determinations	
Day to day precision:	Standard deviation (rel.)	s_w = 10.9%, 6.2% or 10.7%
	Prognostic range	u = 28.0%, 15.9% or 27.5%
	at a spiked concentration of 0.2, 1.0 or 10.0 µg cis-DBCA per litre urine and where n = 6 determinations	

Accuracy:	Recovery rate (rel.) $r = 94\%$
	at a nominal concentration of 1.0 µg cis-DBCA per litre urine and where n = 6 determination within day and from day to day
Detection limit:	0.02 µg cis-DBCA per litre urine
Quantitation limit:	0.06 µg cis-DBCA per litre urine

3-Phenoxybenzoic acid (3-PBA)

Within day precision:	Standard deviation (rel.) $s_w = 7.3\%, 7.0\%$ or 3.6%
	Prognostic range $u = 18.8\%, 18.0\%$ or 9.3%
	at a spiked concentration of 0.2, 1.0 or 10.0 µg 3-PBA per litre urine and where n = 6 determinations
Day to day precision:	Standard deviation (rel.) $s_w = 10.7\%, 7.4\%$ or 11.7%
	Prognostic range $u = 27.5\%, 19.0\%$ or 30.1%
	at a spiked concentration of 0.2, 1.0 or 10.0 µg 3-PBA per litre urine and where n = 6 determinations
Accuracy:	Recovery rate (rel.) $r = 96\%$
	at a nominal concentration of 1.0 µg 3-PBA per litre urine and where n = 6 determinations within day and from day to day
Detection limit:	0.01 µg 3-PBA per litre urine
Quantitation limit:	0.03 µg 3-PBA per litre urine

4-Fluoro-3-phenoxybenzoic acid (FPBA)

Within day precision:	Standard deviation (rel.) $s_w = 1.6\%, 3.6\%$ or 1.2%
	Prognostic range $u = 4.1\%, 9.3\%$ or 3.1%
	at a spiked concentration of 0.2, 1.0 or 10.0 µg FPBA per litre urine and where n = 6 determinations
Day to day precision:	Standard deviation (rel.) $s_w = 6.0\%, 5.9\%$ or 8.8%
	Prognostic range $u = 15.4\%, 15.2\%$ or 22.6%
	at a spiked concentration of 0.2, 1.0 or 10.0 µg FPBA per litre urine and where n = 6 determinations
Accuracy:	Recovery rate (rel.) $r = 93\%$ or 110%
	at a nominal concentration of 1.0 µg FPBA per litre urine and where n = 6 determination within day or from day to day
Detection limit:	0.01 µg FPBA per litre urine
Quantitation limit:	0.03 µg FPBA per litre urine

Pyrethrum and pyrethroids

The pyrethroids and pyrethrum are among the most commonly used insecticides, covering a broad range of applications, for example in agriculture and for interior use, for purposes such as storage and material protection as well as for protection of wood and constructions. Pyrethrum is a natural insecticide obtained from chrysanthemum flowers, whereas pyrethroids are synthetically produced derivatives of the pyrethrins, the native active ingredients of pyrethrum [1, 2]. The production of synthetic pyrethroids aims at obtaining substances that are chemically more stable and effective than pyrethrins but still possess the relatively low toxicity they have in mammals [3]. Although, more than 1000 different pyrethroids have already been chemically synthesised, only a very few are used to a considerable extent. In the USA, permethrin is by far the most commonly used pyrethroid [4].

Within the group of pyrethroids, a MAK value of 10 µg/m³ has been assessed for cyfluthrin only (measured as inhalable fraction) [5]. At present, neither a MAK value nor a BAT value for the other pyrethroids could be established [2].

In humans, pyrethroids such as cyfluthrin, cypermethrin, permethrin and deltamethrin are cleaved by esterases to a wide range of metabolites, including cis- and trans-3-(2,2-dichlorovinyl)-2,2-dimethylcyclopropane carboxylic acid (cis- and trans-DCCA), cis-3-(2,2-dibromovinyl)-2,2-dimethylcyclopropane carboxylic acid (cis-DBCA), 3-phenoxybenzoic acid (3-PBA) and 4-fluoro-3-phenoxybenzoic acid (FPBA).

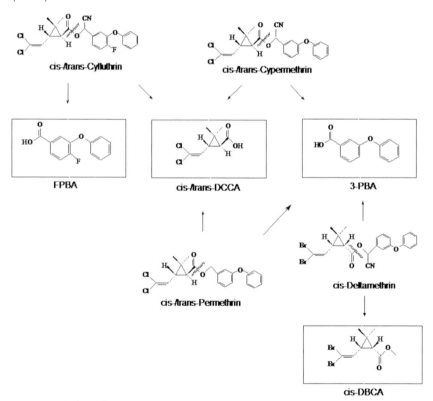

Fig. 1 Metabolism of some selected pyrethroids with the corresponding biomarkers (acc. to [2]).

Cis- and trans-chrysanthemum dicarboxylic acid (cis- and trans-CDCA) are used as specific biomarkers for exposure to pyrethrum, allethrin, resmethrin, phenothrin and tetramethrin [6–9].

For pyrethoids, the liver is the main site of metabolism. The metabolites are exreted in the urine mainly in free or conjugated form (mostly in the form of glucuronides). Depending on the pyrethroid in question, the half-lives are in the range of 4–13 hours [2, 10–14].

Figure 1 shows the metabolism of some selected pyrethroids.

The active ingredients of pyrethrum, the pyrethrins, can be hydrolysed at the central ester bond as well as oxidised by cytochrome P450 monooxygenases [1]. As an example, Figure 2 shows the metabolism of pyrethrin I, for which metabolism studies in humans are already available [7, 15]. After oral intake, the highest trans-CDCA concentrations in urine were found within the first 6 hours. Elimination was complete after 36 hours. Accordingly, the elimination half-life was calculated as 4.2 hours.

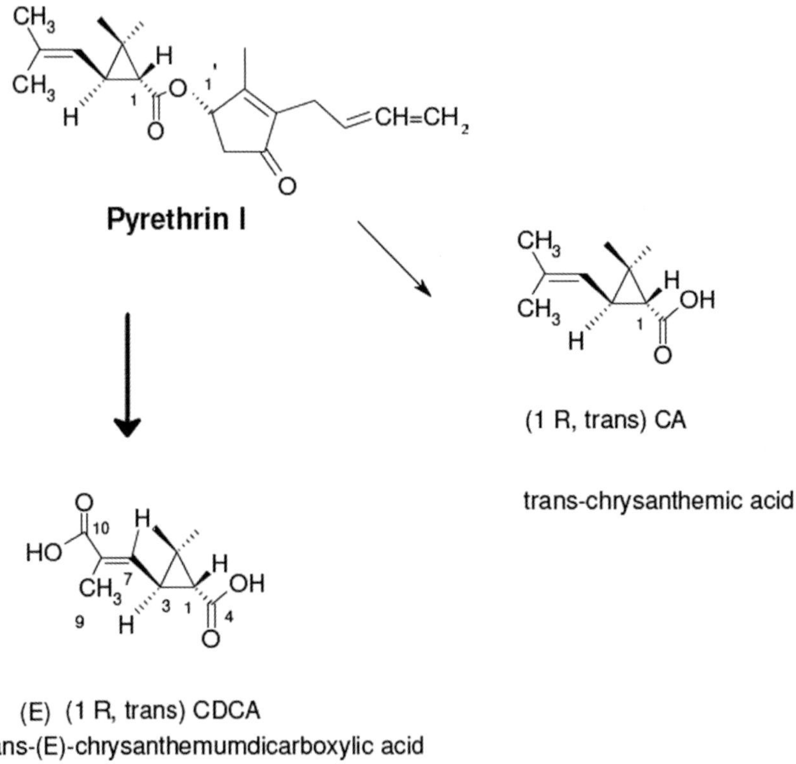

Fig. 2 Metabolism of pyrethrin I (acc. to [2]).

Table 1 Overview of 95th percentiles of the respective metabolite concentrations (µg/L).

References	n	cis-DCCA	trans-DCCA	cis-DBCA	trans-CDCA	3-PBA	FPBA	Sum of metab.
Baker et al. (2004) [19]	45	0.6	0.9	0.1	./.	./.	0.1	1.7
Olsson et al. (2004) [20]	254	0.5*	./.	./.	./.	0.6	./.	1.1
Shan et al. (2004) [21]	1177	0.5	1.5	0.3	./.	./.	0.3	2.6
Class et al. (1990) [22]	57	0.5	0.1	0.1	./.	0.2	0.1	1.0
Kühn (1997) [23]	15	0.95	1.46	0.1	0.12	1.41	0.02	4.1

* given as total DCCA (sum of cis- and trans-DCCA).

For the determination of an internal pyrethroid or pyrethrum exposure, analytical detection of the corresponding metabolites in the urine is recommended for an evaluation of total exposure in the context of occupational and environmental medicine. As pyrethroids and pyrethrum are rapidly metabolised, the urine should ideally be collected directly after the exposure, but at any rate within the first 24 hours after exposure. Otherwise, by biomonitoring, it is not possible to determine the internal pyrethroid or pyrethrum exposure present at the time of the accident. The metabolites found in urine are only suitable as markers of exposure and not of

Table 2 Background levels of the analysed metabolites in 15 urine samples.

Urine sample	trans-CDCA [µg/L]	cis-DCCA [µg/L]	trans-DCCA [µg/L]	cis-DBCA [µg/L]	3-PBA [µg/L]	FPBA [µg/L]
A	<0.05	0.05	0.07	<0.02	0.13	0.021
B	<0.05	0.05	0.09	0.04	0.10	<0.01
C	<0.05	0.06	0.08	0.07	0.11	<0.01
D	<0.05	0.02	0.02	<0.02	0.03	<0.01
E	<0.05	0.03	0.05	<0.02	0.10	<0.01
F	<0.05	0.02	0.03	<0.02	0.05	<0.01
G	0.12	0.07	0.10	0.05	0.16	<0.01
H	0.05	0.21	0.33	0.05	0.47	<0.01
I	<0.05	0.04	0.06	<0.02	0.03	<0.01
J	<0.05	0.36	0.66	0.04	0.70	0.010
K	<0.05	0.07	0.10	<0.02	0.14	<0.01
L*	0.11	2.33	3.35	0.04	3.06	<0.01
M	0.13	0.03	0.05	0.04	0.12	0.024
N	<0.05	0.03	0.05	0.18	0.14	<0.01
O	<0.05	0.07	0.08	0.04	0.13	<0.01

* L: Vegetarian

effects, as several studies have shown that there is no correlation between the occurrence of symptoms and the concentration of metabolites in the urine on the basis of concentration ranges relevant for occupational and environmental medicine [2, 16–18]. For evaluation, the 95th percentiles of the background levels can be used as reference values (see Table 1).

The analyte background levels in urine samples of 15 volunteers analysed with the present method using high resolution sector field mass spectrometer in the NCI mode are listed in Table 2. The observed background levels may possibly be attributed to pyrethrum or pyrethroid residues in fruit and vegetables. This is also supported by the increased levels found in the urine of the volunteer who was adhering to a vegetarian diet.

Authors: *G. Leng, W. Gries*
Examiners: *H.-W. Hoppe, D. Barr*

Pyrethrum and pyrethroid metabolites (after liquid-liquid extraction) in urine

Matrix:	Urine
Hazardous substances:	Pyrethrum and pyrethroids
Analytical principle:	Capillary gas chromatography with mass selective detection (GC-MS)
Completed in:	November 2005

Contents

1	General principles
2	Equipment, chemicals and solutions
2.1	Equipment
2.2	Chemicals
2.3	Solutions
2.4	Internal standard
2.5	Calibration standards
3	Specimen collection and sample preparation
3.1	Specimen collection
3.2	Sample preparation
4	Operational parameters
4.1	Operational parameters for gas chromatography
4.2	Operational parameters for alternative gas chromatography
4.3	Operational parameters for mass spectrometry
4.3.1	Electron impact ionisation (EI)
4.3.2	Negative chemical ionisation (NCI)
4.4	Operational conditions for the alternative GC-MS method
5	Analytical determinations
6	Calibration
7	Calculation of the analytical result
8	Standardisation and quality control
9	Evaluation of the method
9.1	Precision
9.2	Accuracy

9.3	Absolute recovery
9.4	Detection limits and quantitation limits
9.5	Sources of error
10	Discussion of the method
11	References
12	Appendix

1 General principles

After addition of the internal standards, the urine samples are hydrolysed in an acidic environment and the analytes separated from the urine matrix using liquid/liquid extraction with tert-butyl methyl ether. Derivatisation of the carboxylic acids is carried out using hexafluoroisopropanol and diisopropyl carbodiimide. The derivatives are subsequently extracted with isooctane. After capillary gas chromatographic separation, determination of the derivatives is carried out using mass selective detection after electron impact ionisation (EI) or negative chemical ionisation (NCI).

2 Equipment, chemicals and solutions

2.1 Equipment

- Gas chromatograph with split/splitless injector, high-resolution sector field mass spectrometer, autosampler and data processing system
- Capillary gas chromatographic column: length: 30 m; inner diameter: 0.25 mm; stationary phase: 65% diphenylpolysiloxane, 35% polydimethylsiloxane; film thickness: 0.25 µm (e.g. Rtx 65, Restek No. 17023)

Alternative:

- Gas chromatograph with split/splitless injector, quadrupol mass spectrometer with negative chemical ionisation (NCI) unit, autosampler and data processing system
- Capillary gas chromatographic column: length: 60 m; inner diameter: 0.25 mm; stationary phase: 5% phenylpolysiloxane, 95% polydimethylsiloxane; film thickness: 0.25 µm (e.g. Zebron ZB5, Phenomenex No. 7KG-G002-11)

- Various volumetric flasks
- 10 mL Glass centrifuge tubes with teflon-coated screw caps (e.g. Schütt)
- 1.8 mL Crimp cap vials for GC (e.g. Machery-Nagel)
- 200 µL Micro-inserts (e.g. Macherey-Nagel)

- 10 µL and 100 µL Transfer pipettes (e.g. Brand)
- Pasteur pipettes
- Polypropylene containers for urine collection (e.g. Kautex)
- Analytical balance (e.g. Sartorius)
- Drying oven
- Laboratory shaker (e.g. IKA Vibrax VXR)
- Laboratory centrifuge (e.g. Heraeus)
- Evaporation unit (e.g. ReactiVap, Pierce)

2.2 Chemicals

- E-cis-/trans-Chrysanthemum dicarboxylic acid (cis-/trans-CDCA; e.g. Synchem, No. a047)[1)]
- cis-/trans-3-(2,2-Dichlorovinyl)-2,2-dimethylcyclopropane carboxylic acid (cis-/trans-DCCA; e.g. LGC standards, No. CIL-ULM-7303)[1)]
- cis-3-(2,2-Dibromovinyl)-2,2-dimethylcyclopropane carboxylic acid (cis-DBCA; e.g. Ehrenstorfer, No. 12244000)
- 4-Fluoro-3-phenoxybenzoic acid (FPBA; e.g. Promochem, No. CIL-ULM-7391)
- 3-Phenoxybenzoic acid (3-PBA; e.g. Sigma-Aldrich, No. 77708)
- 2-Phenoxybenzoic acid (2-PBA; e.g. Sigma-Aldrich, No. 153176)
- 1,1,1,3,3,3-Hexafluoroisopropanol >99.8% (e.g. Sigma-Aldrich, No. 52517)
- N,N'-Diisopropyl carbodiimide 99% (e.g. Sigma-Aldrich, No. D125407)
- Highly purified water (e.g. Milli-Q-water)
- Acetonitrile anhydrous (e.g. Merck, No. 100004)
- Hydrochloric acid, concentrated, 37% (e.g. Merck, No. 100317)
- Sodium hydrogen carbonate p.a. (e.g. Merck, No. 106329)
- Isooctane for gas chromatography (e.g. Merck, No. 115440)
- tert-Butyl methyl ether ≥ 99.8% (e.g. Sigma-Aldrich, No. 34875)
- Nitrogen 5.0 (e.g. Linde)
- Helium 5.0 (e.g. Linde)
- Ammonia 3.8 (e.g. Linde)
- Methane 4.5 (e.g. Linde)

---

1) Assignment and quantitation of the isomers was carried out by synthesis of the individual standards (Source: Bayer Crop Science). Alternatively, HPLC separation of the isomer mixture is also possible [24].

2.3 Solutions

- 1 M Sodium hydrogen carbonate solution
  42 g sodium hydrogen carbonate are weighed exactly into a 500 mL volumetric flask and dissolved in highly purified water. The flask is then made up to the mark with highly purified water.

2.4 Internal standard

- Stock solution of the internal standard (1 g/L)
  10 mg 2-PBA are weighed exactly into a 10 mL volumetric flask. The flask is then made up to the mark with acetonitrile.
- Working solution of the internal standard (10 mg/L)
  0.1 mL of the stock solution of 2-PBA are placed in a 10 mL volumetric flask which is then made up to the mark with acetonitrile.

The solutions can be kept in the refrigerator (5°C) for at least 6 months.

2.5 Calibration standards

- Stock solutions (1 g/L)
  10 mg each of trans-CDCA, trans-DCCA, cis-DCCA, cis-DBCA, FPBA and 3-PBA are weighed exactly into a 10 mL volumetric flask each, which is made up to the mark with acetonitrile.
- Working solution A (10 mg/L)
  0.1 mL each of the stock solutions are pipetted into a 10 mL volumetric flask, which is made up to the mark with acetonitrile.
- Working solution B (1 mg/L)
  1 mL of working solution A is placed in a 10 mL volumetric flask, which is made up to the mark with acetonitrile.
- Working solution C (100 µg/L)
  0.1 mL of working solution A is placed in a 10 mL volumetric flask, which is made up to the mark with acetonitrile.
- Working solution D (10 µg/L)
  1 mL of working solution C is placed in a 10 mL volumetric flask, which is made up to the mark with acetonitrile.

The solutions can be kept in the refrigerator (5°C) for at least 6 months.

The calibration solutions are prepared in urine. For this purpose, the working solutions A, B, C and D are pipetted into 10 mL screw cap glass tubes according to the pipetting scheme given in Table 3 and diluted with pooled urine. The calibration solutions are then prepared and analysed analogously to the urine samples as described in Section 3 and Section 4.

When stored in a refrigerator (at 5°C), the calibration solutions can be kept for at least 2 weeks.

Table 3 Pipetting scheme for the preparation of calibration solutions.

Volume of working solutions [µL]				Volume of pooled urine [µL]	Concentration of calibration solution [µg/L]
A	B	C	D		
			10	1990	0.05
			20	1980	0.1
			40	1960	0.2
			100	1900	0.5
		20		1980	1.0
		40		1960	2.0
		100		1900	5.0
	20			1980	10.0
	40			1960	20.0
	100			1900	50.0
20				1980	100.0

3 Specimen collection and sample preparation

3.1 Specimen collection

The urine is collected in sealable plastic containers and kept at -20°C until preparation. Stored in this way, the urine is stable for at least 6 months.

3.2 Sample preparation

2 mL urine are transferred to a 10 mL screw cap glass tube using a transfer pipette. After addition of 20 µL of the spiking solution of the internal standard and 0.5 mL concentrated hydrochloric acid, the tube is sealed and heated for at least 2 hours to 100°C in the drying oven.

After equilibration of the solution to room temperature, 4 mL tert-butyl methyl ether are added and the sealed glass tube is subjected to thorough mixing for 10 min on a laboratory shaker. Subsequently, the sample is centrifuged at 2200 g for 5 min. The upper organic phase is transferred with a pasteur pipette to a new 10 mL screw cap glass tube and evaporated to dryness in a stream of nitrogen.

The residue is dissolved in 250 µL acetonitrile, then, 30 µL hexafluoroisopropanol and 15 µL diisopropyl carbodiimide are added for derivatisation. The reaction mixture is gently shaken on a laboratory shaker for 10 min at room temperature. Subsequently, 1 mL 1 M sodium hydrogen carbonate solution is added to the sample solution and briefly mixed. Then, 250 µL isooctane are added and the sample is mixed thoroughly on a laboratory shaker for 10 min and centrifuged (2200 g for 5 min). Subsequently, approx. 150 µL of the isooctane phase are transferred to a GC vial with a micro-insert. An aliquot of 1 µL is used for GC-MS analysis.

The prepared samples can be kept for at least 14 days when stored in the refrigerator (5°C). Longer storage times have not been tested.

4 Operational parameters

4.1 Operational parameters for gas chromatography

GC with high-resolution sector field mass spectrometer

Capillary column:	Material:	Fused Silica
	Stationary phase:	Rtx 65
	Length:	30 m
	Inner diameter:	0.25 mm
	Film thickness:	0.25 µm
Temperature:	Column:	Initial temperature 60°C, 1 min isothermal, then increase at a rate of 8°C/min to 150°C, then increase at 30°C/min to 300°C, 5 min at final temperature
	Injector:	300°C
	Transfer line:	250°C
Carrier gas:	Helium 5.0	120 kPa for 1 min isobar, then increase at 2.5 kPa/min to 140 kPa, isobar up to end of analysis
Split:	40 mL/min	
Septum rinsing:	3 mL/min	
Injection volume:	1 µL	

All other parameters must be optimised in accordance with the manufacturer's instructions.

4.2 Operational parameters for alternative gas chromatography

GC-Quadrupol mass spectrometer

Capillary column:	Material:	Fused Silica
	Stationary phase:	ZB 5
	Length:	60 m
	Inner diameter:	0.25 mm
	Film thickness:	0.25 µm
Temperature:	Column:	Initial temperature 80°C, 1 min isothermal, then increase at a rate of 10°C/min to 200°C, then increase at 30°C/min to 320°C, 5 min at final temperature
	Injector:	300°C
	Transfer line:	280°C
Carrier gas:	Helium 5.0	at a constant flow of 30 cm/s
Split:	40 mL/min	
Septum rinsing:	3 mL/min	
Injection volume:	1 µL	

All other parameters must be optimised in accordance with the manufacturer's instructions.

4.3 Operational parameters for mass spectromety

GC with high-resolution sector field mass spectrometer

4.3.1 Electron impact ionisation (EI)

Ionisation type:	Electron impact ionisation (EI)
Ionisation energy:	70 eV
Source temperature:	250°C
Electron multiplier:	330–350 V

All other parameters must be optimised in accordance with the manufacturer's instructions.

4.3.2 Negative chemical ionisation (NCI)

Ionisation type:	Negative chemical ionisation (NCI)
Reactant gas:	Ammoniac
Ionisation energy:	100 eV
Source temperature:	180°C
Multiplier:	330–350 V

All other parameters must be optimised in accordance with the manufacturer's instructions.

4.4 Operational conditions for the alternative GC-MS method

GC-Quadrupol mass spectrometer

Ionisation type:	Negative chemical ionisation (NCI)
Reactant gas:	Methane (40%)
Ionisation energy:	170 eV
Source temperature:	150°C
Multiplier:	2600 V

All other parameters must be optimised in accordance with the manufacturer's instructions.

5 Analytical determinations

For analytical determination of the urine samples prepared according to Section 3, 1 µL of the isooctane extract is injected into the gas chromatograph and analysed using high-resolution sector field mass spectrometry.

The temporal profiles of the ion traces listed in Table 4 and Table 5 are recorded in the SIR (single ion resolution) mode.

When using the alternative method with Quadrupol-MS, 1 µL of the isooctane extract is injected into the gas chromatograph and analysed using mass spectrometry after negative chemical ionisation (NCI).

The temporal profiles of the ion traces listed in Table 6 are recorded in the SIM (selected ion monitoring) mode.

The retention times given are intended to be a rough guide only. Users of the method must ensure proper separation performance of the capillary column used and the resulting retention behaviour of the analytes.

Table 4 Retention times and ion traces recorded in the EI mode using high resolution sector field mass spectrometry.

Analyte	Retention time [min]	Recorded ion trace [m/z]
trans-CDCA	6.25	331.0769
cis-DCCA	8.25	323.0274
trans-DCCA	8.36	323.0274
cis-DBCA	11.48	368.9750
FPBA	14.42	382.0440
3-PBA	15.04	364.0534
2-PBA (IS)	15.16	364.0534

Table 5 Retention times and ion traces recorded in the NCI mode using high resolution sector field mass spectromety.

Analyte	Retention time [min]	Recorded ion trace [m/z]
trans-CDCA	6.25	330.0691
cis-DCCA	8.25	322.0195
trans-DCCA	8.36	322.0195
cis-DBCA	11.48	365.9690
FPBA	14.42	231.0457
3-PBA	15.04	213.0552
2-PBA (IS)	15.16	213.0552

Table 6 Retention times and ion traces recorded using the alternative quadrupol MS-NCI method.

Analyte	Retention time [min]	Recorded ion trace [m/z]	
		Quantifier	Qualifier
trans-CDCA	12.45	498	330
cis-DCCA	12.80	324	322
trans-DCCA	12.87	324	322
cis-DBCA	15.07	366	81
FPBA	17.27	382	–
3-PBA	17.58	364	213
2-PBA (IS)	17.46	364	–

The chromatogram of a spiked urine sample for each method is given in the Appendix (Figures 4–6).

6 Calibration

The calibration solutions (see Section 2.5) are prepared analogously to the urine samples (see Section 3.2) and analysed using gas chromatography/mass spectrometry as described in Section 4 and Section 5.

The calibration graphs are obtained by plotting the quotients of the peak areas of the analytes and the corresponding internal standards against the spiked concentration used. The linearity of the method was tested and confirmed in the range from 0.05 µg/L to 100 µg/L for high resolution sector field mass spectrometry. For the alternative procedure using Quadrupol-MS, the linearity was established in the range from 0.1 to 10 µg/L. The calibration graphs of the individual analytes are given in the Appendix (see Figure 3).

7 Calculation of the analytical result

To determine the analyte concentration in a sample, the peak area of the analyte is divided by the peak area of the respective internal standard. The quotient thus obtained is inserted into the corresponding calibration graph equation. By dividing the quotient by the slope of the calibration graph, the analyte concentration level of the sample is obtained in µg/L urine. In the case of very high creatinine values (>3 g/L) or if the determined analyte level of the sample is outside the calibration range, the urine sample is diluted and prepared again.

8 Standardisation and quality control

Quality control of the analytical results is carried out as stipulated in the guidelines of the Bundesärztekammer (German Medical Association) [25–27].

To check precision, urine spiked with a known and constant analyte concentration is analysed as control sample within each analytical series. As a material for quality control is not commercially available, it must be prepared in the laboratory by adding a defined quantity of the metabolites to pooled urine. The thus obtained reference urine is divided into aliquots and kept at –20°C. The concentration level of this control material should lie within the relevant concentration range. The nominal value and the tolerance ranges of the quality control material are determined in a pre-analytical period [27]. The determined values of the control samples analysed within each analytical series should lie within the tolerance ranges obtained.

9 Evaluation of the method

9.1 Precision

To determine within day precision, pooled urine was spiked with defined quantities of the metabolites to obtain concentration levels of 0.2 µg/L, 1.0 µg/L and 10 µg/L. These urine samples were prepared and analysed six times in a row. The obtained within day precisions are documented in Table 7.

Table 7 Within day precision for the determination of trans-CDCA, trans- and cis-DCCA, cis-DBCA, 3-PBA and FPBA in urine based on three spiked control urines (n = 6) by analysis using high resolution sector field mass spectrometry.

Analyte	Ionisation mode		Spiked concentration		
			0.2 µg/L	1.0 µg/L	10.0 µg/L
trans-CDCA	EI+	Stand. dev. (rel.) s_w [%]	10.0	5.3	2.7
		Prognostic range u [%]	25.7	13.6	6.9
	NCI	Stand. dev. (rel.) s_w [%]	3.8	4.1	5.1
		Prognostic range u [%]	9.8	10.5	13.1
cis-DCCA	EI+	Stand. dev. (rel.) s_w [%]	4.7	5.2	3.4
		Prognostic range u [%]	12.1	13.4	8.7
	NCI	Stand. dev. (rel.) s_w [%]	7.6	4.5	8.7
		Prognostic range u [%]	19.5	11.6	22.4
trans-DCCA	EI+	Stand. dev. (rel.) s_w [%]	5.2	4.4	2.1
		Prognostic range u [%]	13.4	11.3	5.4
	NCI	Stand. dev. (rel.) s_w [%]	3.8	3.0	7.6
		Prognostic range u [%]	9.8	7.7	19.5
cis-DBCA	EI+	Stand. dev. (rel.) s_w [%]	6.6	3.5	1.9
		Prognostic range u [%]	17.0	9.0	4.9
	NCI	Stand. dev. (rel.) s_w [%]	6.0	4.6	9.3
		Prognostic range u [%]	15.4	11.8	23.9
3-PBA	EI+	Stand. dev. (rel.) s_w [%]	7.8	3.0	1.3
		Prognostic range u [%]	20.0	7.7	3.3
	NCI	Stand. dev. (rel.) s_w [%]	7.3	7.0	3.6
		Prognostic range u [%]	18.8	18.0	9.3
FPBA	EI+	Stand. dev. (rel.) s_w [%]	5.1	3.0	2.1
		Prognostic range u [%]	13.1	7.7	5.4
	NCI	Stand. dev. (rel.) s_w [%]	1.6	3.6	1.2
		Prognostic range u [%]	4.1	9.3	3.1

Furthermore, the day to day precision was determined. For this purpose, the same samples were used. These were prepared and analysed on six different days. The precision data obtained in this manner are shown in Table 8.

Table 8 Day to day precision for the determination of trans-CDCA, cis- and trans-DCCA, cis-DBCA, 3-PBA, FPBA in urine based on three spiked control urines (n = 6) by analysis using high resolution sector field mass spectrometry.

Analyte	Ionisation mode		Spiked analyte level		
			0.2 µg/L	1.0 µg/L	10.0 µg/L
trans-CDCA	EI+	Stand. dev. (rel.) s_w [%]	13.3	3.2	10.7
		Prognostic range u [%]	34.2	8.2	27.5
	NCI	Stand. dev. (rel.) s_w [%]	8.1	6.5	13.1
		Prognostic range u [%]	20.8	16.7	33.7
cis-DCCA	EI+	Stand. dev. (rel.) s_w [%]	13.9	7.0	6.4
		Prognostic range u [%]	35.7	18.0	16.4
	NCI	Stand. dev. (rel.) s_w [%]	11.2	13.3	13.0
		Prognostic range u [%]	28.8	34.2	33.4
trans-DCCA	EI+	Stand. dev. (rel.) s_w [%]	13.0	9.0	6.4
		Prognostic range u [%]	33.4	23.1	16.4
	NCI	Stand. dev. (rel.) s_w [%]	13.4	12.7	10.0
		Prognostic range u [%]	34.3	32.6	25.7
cis-DBCA	EI+	Stand. dev. (rel.) s_w [%]	9.2	10.9	5.4
		Prognostic range u [%]	23.6	28.0	13.9
	NCI	Stand. dev. (rel.) s_w [%]	10.9	6.2	10.7
		Prognostic range u [%]	28.0	15.9	27.5
3-PBA	EI+	Stand. dev. (rel.) s_w [%]	6.9	3.2	1.7
		Prognostic range u [%]	17.7	8.2	4.4
	NCI	Stand. dev. (rel.) s_w [%]	10.7	7.4	11.7
		Prognostic range u [%]	27.5	19.0	30.1
FPBA	EI+	Stand. dev. (rel.) s_w [%]	6.7	4.7	2.5
		Prognostic range u [%]	17.2	12.1	6.4
	NCI	Stand. dev. (rel.) s_w [%]	6.0	5.9	8.8
		Prognostic range u [%]	15.4	15.2	22.6

To determine the within day precision of the alternative method using Quadrupol mass spectrometry (see Section 4.2 and Section 4.4), pooled urine was spiked to obtain two concentrations (1.0 µg/L and 10 µg/L) and analysed six times in a row. The obtained within day precision data are documented in Table 9.

Table 9 Within day precision for the determination of trans-CDCA, cis- and trans-DCCA, cis-DBCA, 3-PBA, FPBA in urine based on two spiked concentrations (n = 6) by analysis using the alternative procedure (Quadrupol-MS).

Analyte	Ionisation mode		Spiked concentration	
			1.0 µg/L	10.0 µg/L
trans-CDCA	NCI	Standard deviation (rel.) s_w [%]	8.0	8.7
		Prognostic range u [%]	20.6	22.4
cis-DCCA	NCI	Standard deviation (rel.) s_w [%]	6.1	8.1
		Prognostic range u [%]	15.7	20.8
trans-DCCA	NCI	Standard deviation (rel.) s_w [%]	9.1	8.2
		Prognostic range u [%]	23.4	21.1
cis-DBCA	NCI	Standard deviation (rel.) s_w [%]	8.2	8.7
		Prognostic range u [%]	21.1	22.4
3-PBA	NCI	Standard deviation (rel.) s_w [%]	10.3	9.6
		Prognostic range u [%]	26.5	24.7
FPBA	NCI	Standard deviation (rel.) s_w [%]	7.8	8.7
		Prognostic range u [%]	20.0	22.3

Furthermore, the day to day precision was also determined for the alternative procedure using Quadrupol-MS (see Section 4.2 and Section 4.4). For this purpose, the two spiked urine samples were also used. These were prepared and analysed on six different days. The precision data obtained in this manner are shown in Table 10.

Table 10 Day to day precision for the determination of trans-CDCA, trans- and cis-DCCA, cis-DBCA, 3-PBA and FPBA in urine based on two spiked control urine samples (n = 6) by analysis using the alternative procedure (Quadrupol-MS).

Analyte	Ionisation mode		Spiked concentration	
			1.0 µg/L	10.0 µg/L
trans-CDCA	NCI	Standard deviation (rel.) s_w [%]	12.9	10.2
		Prognostic range u [%]	33.2	26.2
cis-DCCA	NCI	Standard deviation (rel.) s_w [%]	10.2	9.4
		Prognostic range u [%]	26.2	24.2
trans-DCCA	NCI	Standard deviation (rel.) s_w [%]	7.1	9.2
		Prognostic range u [%]	18.2	23.6
cis-DBCA	NCI	Standard deviation (rel.) s_w [%]	6.7	7.9
		Prognostic range u [%]	17.2	20.3
3-PBA	NCI	Standard deviation (rel.) s_w [%]	14.0	9.7
		Prognostic range u [%]	36.0	24.9
FPBA	NCI	Standard deviation (rel.) s_w [%]	7.4	8.9
		Prognostic range u [%]	19.0	22.9

9.2 Accuracy

To determine the accuracy of the method, recovery experiments were carried out. For this purpose, pooled urine was spiked with defined concentrations (0.2 µg/L, 1.0 µg/L and 10 µg/L) of the metabolites to be investigated. The samples were processed and analysed six times as described in Section 3.2. The relative recovery rates obtained are summarised in Table 11.

Recovery experiments were also carried out to determine the accuracy of the alternative method using Quadrupol-MS (see Section 4.2 and Section 4.4). For this

Table 11 Relative recovery rates for trans-CDCA, cis- and trans-DCCA, cis-DBCA, 3-PBA and FPBA in urine based on three spiked control urine samples with determination within day as well as from day to day (n = 6) and analysis using high-resolution sector field MS (rel. rec. = mean relative recovery rate).

Analyte	Ionisation mode		Spiked concentration		
			0.2 µg/L	1.0 µg/L	10.0 µg/L
trans-CDCA	EI+	Rel. rec. r [%] within day	90	95	102
		Rel. rec. r [%] day to day	86	103	88
	NCI	Rel. rec. r [%] within day	114	90	102
		Rel. rec. r [%] day to day	103	107	105
cis-DCCA	EI+	Rel. rec. r [%] within day	106	104	97
		Rel. rec. r [%] day to day	104	87	94
	NCI	Rel. rec. r [%] within day	104	94	87
		Rel. rec. r [%] day to day	90	90	99
trans-DCCA	EI+	Rel. rec. r [%] within day	108	97	97
		Rel. rec. r [%] day to day	105	86	93
	NCI	Rel. rec. r [%] within day	96	106	97
		Rel. rec. r [%] day to day	83	102	93
cis-DBCA	EI+	Rel. rec. r [%] within day	94	102	99
		Rel. rec. r [%] day to day	95	96	100
	NCI	Rel. rec. r [%] within day	94	94	74
		Rel. rec. r [%] day to day	95	94	97
3-PBA	EI+	Rel. rec. r [%] within day	101	100	101
		Rel. rec. r [%] day to day	99	99	92
	NCI	Rel. rec. r [%] within day	83	96	105
		Rel. rec. r [%] day to day	76	96	101
FPBA	EI+	Rel. rec. r [%] within day	101	104	101
		Rel. rec. r [%] day to day	104	101	101
	NCI	Rel. rec. r [%] within day	94	93	103
		Rel. rec. r [%] day to day	91	110	99

purpose, pooled urine was spiked with defined concentrations (1.0 µg/L and 10 µg/L) of the metabolites. The samples were processed and analysed six times as described in Section 3.2. The relative recovery rates obtained are summarised in Table 12.

Table 12 Relative recovery rates for trans-CDCA, cis- and trans-DCCA, cis-DBCA, 3-PBA and FPBA in urine based on two spiked control urine samples with determination within day as well as from day to day (n = 6) and analysis using the alternative procedure (Quadrupol-MS).

Analyte	Ionisation mode		Spiked concentration	
			1.0 µg/L	10.0 µg/L
trans-CDCA	NCI	Rel. rec. r [%] within day	92	92
		Rel. rec. r [%] from day to day	87	104
cis-DCCA	NCI	Rel. rec. r [%] within day	89	94
		Rel. rec. r [%] from day to day	89	107
trans-DCCA	NCI	Rel. rec. r [%] within day	99	95
		Rel. rec. r [%] from day to day	101	95
cis-DBCA	NCI	Rel. rec. r [%] within day	89	96
		Rel. rec. r [%] from day to day	106	95
3-PBA	NCI	Rel. rec. r [%] within day	89	103
		Rel. rec. r [%] from day to day	101	105
FPBA	NCI	Rel. rec. r [%] within day	85	94
		Rel. rec. r [%] from day to day	107	94

9.3 Absolute recovery

Additionally, the examiner of the method studied the absolute recovery and the analyte losses due to sample preparation, respectively. For this purpose, urine samples were spiked with the metabolites before and after liquid-liquid extraction and the absolute recovery was calculated. The results are summarised in Table 13.

Table 13 Absolute recovery of the analytes trans-CDCA, cis- and trans-DCCA, cis-DBCA, 3-PBA and FPBA in urine after liquid-liquid extraction with four urine samples spiked with different concentration levels (n = 6).

Analyte	Mean absolute recovery [%] at a spiked analyte level of			
	0.2 µg/L	1 µg/L	10 µg/L	100 µg/L
trans-CDCA	85	73	77	82
cis-DCCA	91	68	62	64
trans-DCCA	80	71	68	70
cis-DBCA	–*	–*	75	72
3-PBA	99	89	80	89
FPBA	99	88	81	87

* Value could not be determined reliably.

9.4 Detection limits and quantitation limits

The detection limit was estimated on the basis of a signal to noise ratio of 3 : 1. The quantitation limit was calculated on the basis of a signal to noise ratio of 9 : 1. The values thus obtained by the different method alternatives are summarised in Table 14.

Table 14 Detection limits and quantitation limits for trans-CDCA, cis- and trans-DCCA, cis-DBCA, 3-PBA and FPBA in urine for the alternative methods (LOD: limit of detection, LOQ: limit of quantitation).

Analyte	High resolution sector field MS				Quadrupol-MS	
	EI+ mode		NCI mode		NCI mode	
	LOD [µg/L]	LOQ [µg/L]	LOD [µg/L]	LOQ [µg/L]	LOD [µg/L]	LOQ [µg/L]
trans-CDCA	0.05	0.15	0.05	0.15	0.1	0.3
trans-DCCA	0.05	0.15	0.02	0.06	0.1	0.3
cis-DCCA	0.05	0.15	0.02	0.06	0.1	0.3
cis-DBCA	0.1	0.3	0.02	0.06	0.1	0.3
3-PBA	0.02	0.06	0.01	0.03	0.2	0.6
FPBA	0.02	0.06	0.01	0.03	0.1	0.3

9.5 Sources of error

Analysis in the EI mode is recommended for routine applications, when using a high resolution sector field mass spectrometer. This is because the EI source is not so rapidly contaminated and can thus be used longer before cleaning is necessary. However, analysis in the NCI mode is more sensitive as regards most analytes and therefore suitable for confirmation analyses or for determinations at ultra trace level. But, analysis using NCI mode is generally more complex and may result in higher contamination of the ion source.

When using a quadrupol mass spectrometer (alternative method), determination with sufficient sensitivity is only possible if the ionisation takes place in the NCI mode. However, interfering signals on the ion trace $m/z = 322$ occur in a number of urine samples which may lead to a signal overlap with cis- and trans-DCCA. In such cases, quantitation of these analytes via the mass trace $m/z = 324$ may be advantageous. Moreover, this approach does not involve a significant loss in sensitivity.

A high dosage of the calibration gas perfluorokerosine (PFK) may lead to a quenching of the mass fragments $m/z = 366.9768$ and $m/z = 368.9748$ and, as a result, to sensitivity losses of cis-DBCA, when using a high resolution sector field mass spectrometer. In such cases, the ion trace $m/z = 252.9045$ can be used as an alternative mass trace for quantitation.

The internal standard 2-PBA compensates almost all analytical deviations and is therefore well suited as internal standard for the investigated pyrethrum and pyrethroid metabolites.

10 Discussion of the method

With the described procedure, both, the metabolites formed from natural pyrethrum and those from the synthetic pyrethroids can be analysed within one analytical procedure.

By the use of sensitive GC-MS technology in the form of high resolution sector field mass spectrometry or quadrupol mass spectrometry with negative chemical ionisation, the high sensitivity of the method is made possible.

When using sector field mass spectrometry analysis, the EI mode is recommended for routine measurements. Laboratories without this equipment may alternatively use a Quadrupol-MS system in the NCI mode, for which the present method was also validated.

For method validation, apart from 2-PBA, the commercially not available internal standard $^{13}C_4,d_3$-cis-DCCA was also tested. No deviations to the here described method were observed when using sector field mass spectrometry. Analysis using Quadrupol-MS and the above mentioned isotopically labelled IS did, however, produce lower variation coefficients compared with 2-PBA. Nevertheless, the basic functionality of 2-PBA as internal standard is not challenged as the method validation yielded absolute recovery rates of approx. 80% for all analytes with liquid-liquid extraction. The high and similar recovery rates of all analytes confirm the good suitability of the applied sample preparation procedure as well as the usefulness of the application of only one internal standard to compensate for analyte losses.

The present method involves a rapid and simple sample preparation so that time-consuming clean-up steps are not necessary. Still, by dilution of the sample solutions, a good performance of the analysis system is ensured. When using a Quadrupol-MS system in the NCI mode, a very good sensitivity is obtained for all investigated metabolites, which is quite comparable with that of a high resolution sector field MS system in the EI mode. Using the alternative method with an OPTIMA-5 column (30 m x 0.5 mm x 0.25 µm) for analyte separation, the examiner of the method was able to obtain somewhat lower detection limits for 3-PBA and FPBA of 0.05 µg/L each. Using this column, a more effective separation of interfering compounds could be observed. However, with all method variants, precision and accuracy were always highly satisfactory.

In principle, determination of cis-chrysanthemum dicarboxylic acid (cis-CDCA) is also possible with the present method. But, according to investigations by the authors, the formation of this metabolite seems to play no role in human metabolism [7]. A validation of this parameter was therefore not carried out.

Instruments used:

- Gas chromatograph (HP 5890 II Plus, Agilent, USA) with split/splitless injector and autosampler (CTC A 200 S, Analytics, Switzerland) in conjunction with a sector field mass spectrometer (Autospec Ultima, Micromass/Waters). As software, Opus 3.6 was used for analysis and evaluation.
- For the alternative method a GC Quadrupol MS system (MSD 5973, Agilent, USA) with negative chemical ionisation was used.

11 References

1. H. Greim (ed.): Pyrethrum. Gesundheitsschädliche Arbeitsstoffe, Toxikologisch-arbeitsmedizinische Begründung von MAK-Werten. 45th issue, Wiley-VCH Verlag, Weinheim (2008).
2. H. Greim and H. Drexler (eds.): Pyrethroide. Biologische Arbeitsstoff-Toleranz-Werte (BAT-Werte) und Expositionsäquivalente für krebserzeugende Arbeitsstoffe (EKA) und Biologische Leitwerte (BLW). 15th issue, Wiley-VCH-Verlag, Weinheim (2008).
3. D.M. Soderlund, J.M. Clark, L.P. Sheets, L.S. Mullin, V.J. Piccirillo, D. Sargent, J.T. Stevens, M.L. Weiner: Mechanisms of pyrethroid neurotoxicity: implications for cumulative risk assessment. Toxicology 171, 3–59 (2002).
4. Agency for Toxic Substances and Disease Registry: Toxicological profile for pyrethrins and pyrethroids. US Departement of Health and Human Services, Atlanta (2003).
5. H. Greim (ed.): Cyfluthrin. The MAK Collection Part I: MAK Value Documentations. Volume 23, Wiley-VCH Verlag, Weinheim (2007).
6. G. Leng, K.-H. Kühn, B. Wieseler, H. Idel: Metabolism of (S)-bioallethrin and related compounds in humans. Toxicol. Lett. 107, 109–121 (1999).
7. G. Leng, W. Gries, S. Selim: Biomarker of pyrethrum exposure. Toxicol. Lett. 162, 195–201 (2006).
8. G. Leng, K.-H. Kühn, H. Idel: Biological monitoring of pyrethroids in blood and pyrethroid metabolites in urine: applications and limitations. Sci. Total Environ. 199, 173–181 (1997).
9. J. Angerer and K.-H. Schaller (eds.): Pyrethroid Metabolites. Analyses of hazardous substances in biological materials. Volume 6, Wiley-VCH-Verlag, Weinheim, 231–254 (1999).
10. F. He, S. Wang, L. Liu, S. Chen, Z. Zhang, J. Sun: Clinical manifestations of acute pyrethroid poisoning. Arch. Toxicol. 63, 54–58 (1989).
11. C.V. Eadsforth and M.K. Baldwin: Human dose-excretion studies with the pyrethroid insecticide cypermethrin. Xenobiotica 13, 67–72 (1983).
12. C.V. Eadsforth, P.C. Bragt, N.J. van Sittert: Human dose-excretion studies with pyrethroid insecticides cypermethrin and alphacypermethrin: relevance for biological monitoring. Xenobiotica 18, 603–614 (1988).
13. B.H. Woollen, J.R. Marsh, W.J.D. Laird, J.E. Lesser: The metabolism of cypermethrin in man: differences in urinary metabolite profiles following oral and dermal administration. Xenobiotica 22, 893–991 (1992).
14. G. Leng, A. Leng, K.-H. Kühn, J. Lewalter, J. Pauluhn: Human dose excretion studies with the pyrethroid insecticide cyfluthrin: urinary metabolite profile following inhalation. Xenobiotica 27, 1272–1283 (1997).

15 J. Angerer and A. Ritter: Determination of metabolites of pyrethroids in human urine using solid-phase extraction and gas chromatography–mass spectrometry. J. Chromatogr. B 695, 217–226 (1997).
16 K.-H. Kühn, G. Leng, K.A. Bucholski, L. Dunemann, H. Idel: Determination of pyrethroid metabolites in human urine by capillary gas chromatography-mass spectrometry. Chromatographia 43, 285–292 (1996).
17 T. Schettgen, H.M. Koch, H. Drexler, J. Angerer: New gas chromatographic-mass spectrometric method for the determination of urinary pyrethroid metabolites in environmental medicine. J. Chromatogr. B 778, 121–130 (2002).
18 F.J. Arrebola, J.L. Martinez-Vidal, A. Fernández-Gutiérrez, M.H. Akhtar: Monitoring of pyrethroid metabolites in human urine using solid-phase extraction followed by gas chromatography-tandem mass spectrometry, Anal. Chim. Acta 401, 45–54 (1999).
19 S.E. Baker, A.O. Olsson, D.B. Barr: Isotope dilution high-performance liquid chromatography-tandem mass spectrometry method for quantifying urinary metabolites of synthetic pyrethroid insecticides. Arch. Environ. Contam. Toxicol. 46, 281–288 (2004).
20 A.O. Olsson, S.E. Baker, J.V. Nguyen, L.C. Romanoff, S.O. Udunka, R.D. Walker, K.L. Flemmen, D.B. Barr: A liquid chromatography-tandem mass spectrometry multiresidue method for quantification of specific metabolites of organophosphorus pesticides, synthetic pyrethroids, selected herbicides, and deet in human urine. Anal. Chem. 76, 2453–2461 (2004).
21 G. Shan, H. Huang, D.W. Stoutamire, S.J. Gee, G. Leng, B.D. Hammock: A sensitive class specific immunoassay for the detection of pyrethroid metabolites in human urine. Chem. Res. Toxicol. 17, 218–225 (2004).
22 T.J. Class, T. Ando, J.E. Casida: Pyrethroid metabolism: microsomal oxidase metabolites of (S)-bioallethrin and the six natural pyrethrins. J. Agr. Food Chem. 38, 529–537 (1990).
23 K.H. Kühn: Bestimmung von Pyrethroiden und ihren Metaboliten in Blut und Urin mittels GC/MS und GC/ECD. Thesis Clausthal, Shaker Verlag Aachen (1997).
24 L. Elflein, E. Berger-Preiss, A. Preiss, M. Elend, K. Levsen, G. Wünsch: Human biomonitoring of pyrethrum and pyrethroid insecticides used indoors: determination of the metabolites E-cis/trans-chrysanthemumdicarboxylic acid in human urine by gas chromatography–mass spectrometry with negative chemical ionization. J. Chromatogr. B 795, 195–207 (2003).
25 Bundesärztekammer: Qualitätssicherung der quantitativen Bestimmungen im Laboratorium. Neue Richtlinien der Bundesärztekammer. Dt. Ärztebl. 85, A699–A712 (1988).
26 Bundesärztekammer: Ergänzung der "Richtlinien der Bundesärztekammer zur Qualitätssicherung in medizinischen Laboratorien". Dt. Ärztebl. 91, C159–C161 (1994).
27 J. Angerer, T. Göen, G. Lehnert: Mindestanforderungen an die Qualität von umweltmedizinisch-toxikologischen Analysen. Umweltmed. Forsch. Prax. 3, 307–312 (1998).

Authors: *G. Leng, W. Gries*
Examiners: *H.-W. Hoppe, D. Barr*

12   Appendix

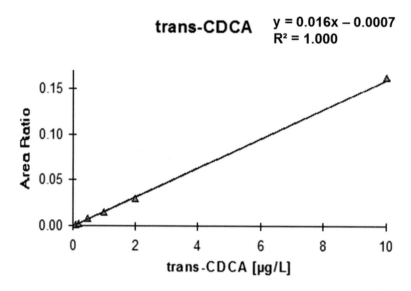

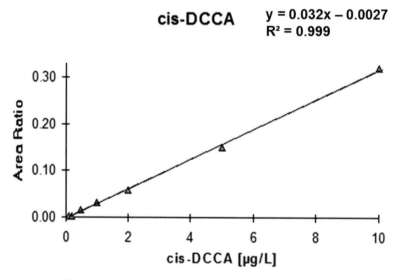

Fig. 3   a) Calibration graphs of the analysed metabolites in urine in the concentration range from 0.1 to 10 µg/L: trans-CDCA and cis-DCCA.

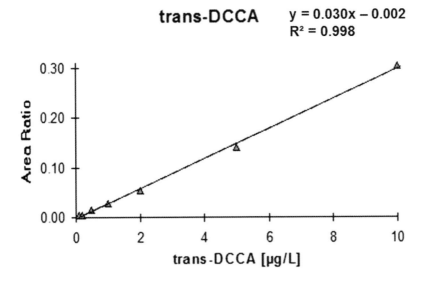
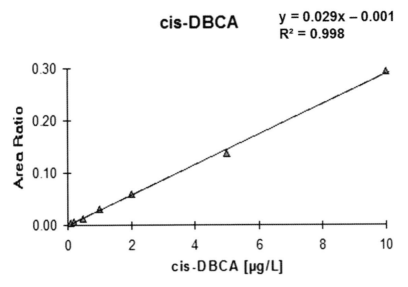

Fig. 3 b) Calibration graphs of the analysed metabolites in urine in the concentration range from 0.1 to 10 µg/L: trans-DCCA and cis-DBCA.

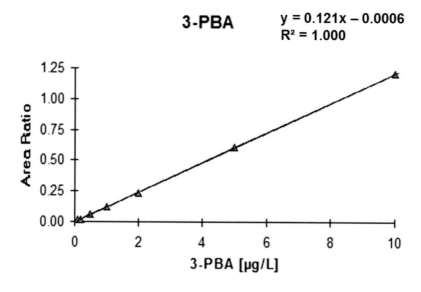

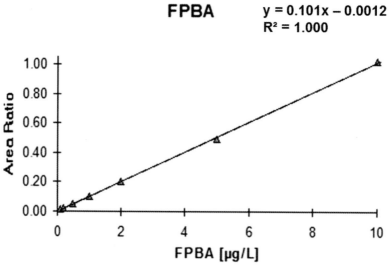

Fig. 3 c) Calibration graphs of the analysed metabolites in urine in the concentration range from 0.1 to 10 µg/L: 3-PBA and FPBA.

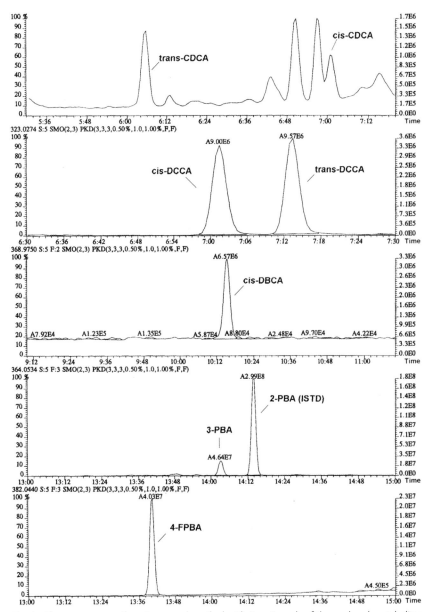

Fig. 4 Chromatograms of a urine sample spiked with 1 µg/L each of the analysed metabolites using high resolution sector field MS in the EI mode.

246 | *Analytical Methods*

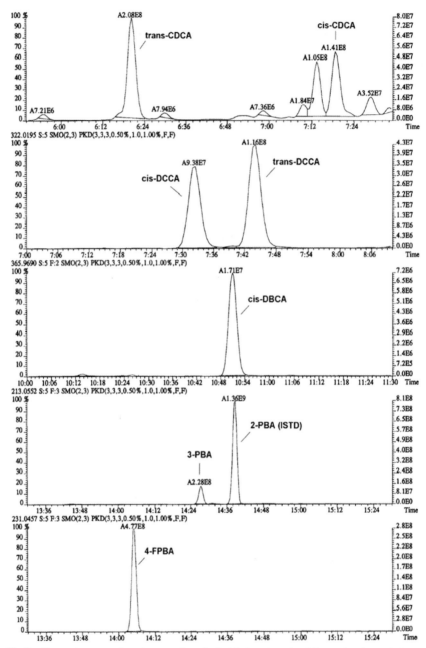

Fig. 5 Chromatograms of a urine sample spiked with 1 μg/L each of the analysed metabolites using high resolution sector field MS in the NCI mode.

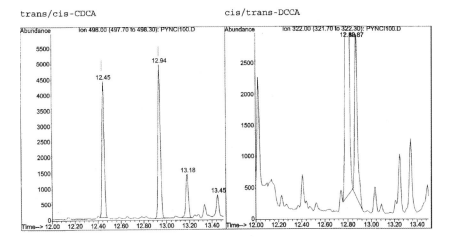

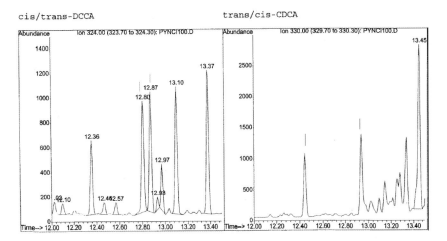

Fig. 6 a) Chromatograms of a urine sample spiked with 1 µg/L each of the analysed metabolites using Quadrupol MS (NCI mode).

248 | Analytical Methods

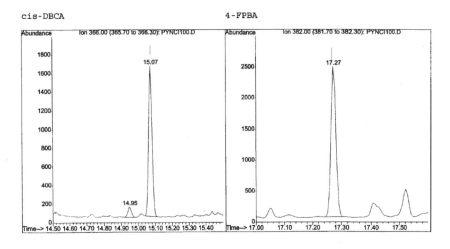

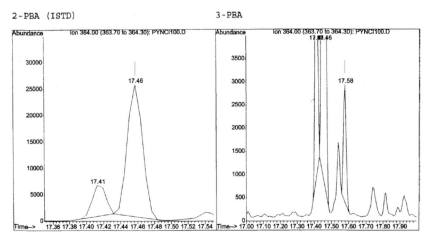

Fig. 6 b) Chromatograms of a urine sample spiked with 1 µg/L each of the analysed metabolites using Quadrupol MS (NCI mode).

Selenium in serum

Matrix:	Serum or plasma
Hazardous substances:	Selenium and its compounds
Analytical principle:	Graphite furnace AAS
Completed in:	August 2008

Overview of parameters that can be determined with this method and the corresponding hazardous substances:

Hazardous substance	CAS	Parameter	CAS
Selenium	7782-49-2		
Inorganic selenium compounds	–	Selenium	7782-49-2
Organic selenium compounds	–		

Summary

This analytical method allows the quantitative determination of the selenium concentration in serum or plasma using graphite furnace atomic absorption spectrometry with Zeeman background correction. The serum or plasma samples are diluted in a ratio of 1/10 (v:v) with Triton™ X-100 solution and a Pd-Mg matrix modifier and are injected directly into the graphite furnace of the AAS (by use of an autosampler). Quantitative determination is carried out after calibration in matrix, leading to a strictly linear correlation between the measured extinction and the concentration level of selenium.

The method is very sensitive and applicable in routine laboratories with high sample throughput. The detection limit is 3 µg selenium per litre serum, thus selenium can be determined reliably even in the case of extreme selenium deficiency.

Selenium

Within day precision:	Standard deviation (rel.)	s_w = 3.0% or 2.1%
	Prognostic range	u = 6.7% or 4.7%
	at a concentration of 81 µg or 136 µg selenium per litre serum and where n = 10 determinations	
Day to day precision:	Standard deviation (rel.)	s_w = 4.8% or 4.2%
	Prognostic range	u = 10.0% or 8.8%
	at a concentration of 81 µg or 136 µg selenium per litre serum and where n = 20 determinations	
Accuracy:	Recovery rate (rel.)	r = 97%
	at a nominal concentration of 100 µg selenium per litre serum and where n = 5 determinations	
Detection limit:	3 µg selenium per litre serum	
Quantitation limit:	9 µg selenium per litre serum	

Selenium

No other element has undergone such a dramatic change in its physiological and toxicological assessment over the years as selenium. Up to the 1930s, for example, selenium (Se), like arsenic, was considered to be extremely toxic or carcinogenic [1, 2]. During the 20th century, however, this assessment has changed continuously. Growing evidence for the essentiality of selenium has been reported [3] and in 1973 selenium was described as a part of glutathione peroxidase [4]. Today, selenium is considered to be a trace element of extraordinary importance, especially as essential component in the diet and as a dietary supplement in the prevention and therapy of various diseases, including cancer. This is based on the fact that it is part of specific selenoproteins and/or selenoenzymes, as for example glutathione peroxidase, iodothyronine deiodase, thioreduxin reductase or selenoprotein P [4–7].

Selenium is No. 69 of the elements in the earth's crust and thus belongs to the rarer elements. The burning of fossil fuels and volcanic activity are responsible for the distribution of selenium in the water or the atmosphere. Both, inorganic and organic selenium forms are found in nature. Whereas the inorganic species predominate in the inanimate nature or in industrial processes (e.g. selenite, selenate, elemental selenium, hydrogen selenide), the organic species (e.g. selenomethionine, selenocysteine) predominate in animate nature.

Food consumption is the most important route for human exposure to selenium, whereby the selenium content of foods can vary considerably according to the foodstuffs' origin [8]. The content of selenium in food depends on the selenium content of the soil where plants are grown or animals raised and differs largely [9]. Food of animal origin, in general, contains more selenium than plant based food. In drinking water selenium concentrations are in the low µg/L range. The German Drinking Water Ordinance sets a legal limit of 10 µg selenium per litre drinking water [10].

Occupational exposure to selenium and its inorganic compounds tends to be rare and is described in single cases [11, 12]. Detailed information on the toxicity of selenium and its inorganic compounds is published in the 1999 and 2011 MAK documentations [11, 12]. The maximum workplace concentration (MAK value) for selenium and its inorganic compounds is set at 0.02 mg/m^3 [12].

In addition to the quantification of occupational exposure, the determination of selenium levels is nowadays important to diagnose selenium deficiency. Depending on the objectives, selenium determination can be carried out in blood, urine, serum or plasma. Selenium levels in whole blood allow a statement on the selenium status as a long-term parameter [29]. Although its concentration in urine

Table 1 Concentration range of selenium in serum or plasma in different countries.

Country/Region	Group/Normal values	Reference
Germany	Adults 74–139 µg/L	[13, 14]
	Babies 1–4 months 18–64 µg/L	
	Babies 5–12 months 32–101 µg/L	
	Infants 58–116 µg/L	
	School children 69–121 µg/L	
Germany	Adults 50–120 µg/L	[15]
	Children 0–1 year 33–71 µg/L	
	Children 2–5 years 32–84 µg/L	
	Children 5–10 years 41–74 µg/L	
	Children 10–16 years 40–82 µg/L	
Germany	Children (average age 10.3 years, n = 1918)	[16]
	33–98 µg/L (range),	
	55–72 µg/L (mean values of individual subcollectives)	
Germany	Children 1–5 years (n = 221)	[17]
	74 µg/L (mean value), 41–116 µg/L (5–95th percentile)	
	Children 6–18 years (n = 623)	
	79 µg/L (mean value), 48–116 µg/L (5–95th percentile)	
Canada	Babies 45–104 µg/L (range) (n = 20)	[18]
	Children 1–5 years 99–142 µg/L (range)(n = 20)	
	Children 6–9 years 111–164 µg/L (range) (n = 20)	
	above 10 years and adults 102–205 µg/L (range) (n = 57)	
Iran/Teheran	Women >16 years 67–121 µg/L (5–95th percentile) (n = 24)	[19]
	Men >16 years 79–126 µg/L (5–95th percentile) (n = 106)	
Kuwait	Adults 88 µg/L (mean value) (n = 379)	[20]
USA	Adults >20 years 126 µg/L (mean value) (n = 7497)	[21]
USA	Children 3–11 years 112 µg/L (geometric mean value),	[22]
	93–130 µg/L (10–90th percentile) (n = 1186)	
Japan, Tokyo	Adults 146 µg/L (mean value) (n = 118)	[23]
Taiwan	Adults 111 µg/L (mean value), 41–186 µg/L (range) (n = 2755)	[24]
India	Adults 100 µg/L (mean value), 36–186 µg/L (range) (n = 201)	[25]
Singapore	Adults 122 µg/L (mean value) (n = 244)	[26]
Switzerland	Adults 98 µg/L (mean value), (n = 1847)	[27]
Spain	Adults 111 µg/L (mean value), 60–106 µg/L (range), (n = 150)	[28]

gives no information on the functional selenium status, it reflects the current selenium intake [29]. For this reason an occupational medical determination of selenium levels in urine makes sense. However, most frequently serum or plasma is used for biomonitoring, as the selenium levels in these matrices represent the current selenium status. Moreover, analytical selenium determination in serum or plasma is less problematic and more reliable than in blood or urine. This is why selenium determination in serum was used for deriving a BAT value (*Biologischer Arbeitsstoff-Toleranz-Wert*). This value is set at 150 µg selenium per litre serum [30].

Table 1 shows that selenium concentration in serum or plasma depends greatly on the region where people live. Considerable regional differences even exist within the individual countries [22]. As a rule, children show lower selenium levels than adults [13]. Selenium can be detected both in plasma and in the erythrocytes. Due to the slightly higher selenium content in the erythrocytes the levels in the whole blood are about 1.3 times higher than those found in plasma [31].

Author: *P. Heitland*
Examiner: *B. Michalke*

Selenium in serum

Matrix:	Serum or plasma
Hazardous substances:	Selenium and its compounds
Analytical principle:	Graphite furnace AAS
Completed in:	August 2008

Contents

1	General principles
2	Equipment, chemicals and solutions
2.1	Equipment
2.2	Chemicals
2.3	Solutions
2.4	Calibration standards
3	Specimen collection and sample preparation
3.1	Specimen collection
3.2	Sample preparation
4	Operational parameters
5	Analytical determination
6	Calibration
7	Calculation of the analytical result
8	Standardisation and quality control
9	Evaluation of the method
9.1	Precision
9.2	Accuracy
9.3	Detection limit and quantitation limit
9.4	Sources of error
10	Discussion of the method
11	References
12	Appendix

1 General principles

Selenium is determined using graphite furnace atomic absorption spectrometry with Zeeman background correction. The serum/plasma samples are diluted in a ratio of 1/10 (v:v) with Triton™X-100 solution and a Pd-Mg matrix modifier and are injected directly into the graphite furnace of the AAS spectrometer. Quantitative determination is carried out after calibration in matrix, resulting in a linear relationship between the measured extinction and the spiked concentration levels of selenium. The method is very sensitive and applicable in routine laboratories with high sample throughput. The detection limit is 3 µg Se per litre serum, thus selenium can be determined reliably even in the case of extreme selenium deficiency.

2 Equipment, chemicals and solutions

2.1 Equipment

- Graphite furnace atomic absorption spectrometer with Zeeman background correction and autosampler (e.g. Thermo Elemental)
- Stroke pipettes with adjustable volume between 10–100 µL or 100–1000 µL with suitable pipette tips (e.g. Eppendorf)
- 1000 mL Polyethylene flask with bottletop dispenser (adjustable between 0.5 and 5 mL) (e.g. Dispensette™, Brand)
- 10 mL, 100 mL and 1000 mL Volumetric flasks (e.g. Schott)
- 10 mL Polypropylene vials for autosampler use (e.g. Sarstedt)
- Laboratory shaker (e.g. Reax 2000; Heidolph Instruments GmbH)
- 1.5 mL Sample vials (e.g. Omnilab)
- Lithium-Heparine monovette™ for plasma sampling (e.g. Sarstedt)
- Neutral-S monovette™ for serum sampling (e.g. Sarstedt)

2.2 Chemicals

- Selenium standard solution CertiPUR™; 1000 mg/L Se (e.g. Merck, No. 119796)
- Analytical quality control material (e.g. Seronorm™ Trace Element Serum)
- Nitric acid 65% Suprapur™ (e.g. Merck, No. 100441)
- Magnesium matrix modifier 10 g/L (e.g. Merck, No. 105813)
- Palladium matrix modifier 10 g/L (e.g. Merck, No. 107289)
- Albumin fraction V (from bovine serum) (e.g. Merck, No. 112018)
- Triton™X-100 (e.g. Sigma-Aldrich, No. X100)
- Deionised water

- Argon 5.0 (e.g. Linde)
- Synthetic air (e.g. Linde)

2.3 Solutions

- Dilution solution
  1 mL Triton™X-100 is pipetted into a 1000 mL volumetric flask and dissolved in deionised water. The flask is then made up to the mark with deionised water.
- Matrix modifier solution
  10 mL Pd matrix modifier (10 g/L Pd), 5 mL Mg matrix modifier (10 g/L Mg), 0.8 mL Triton™X-100 and 0.5 mL 65% nitric acid are pipetted into a 100 mL volumetric flask. The flask is then made up to the mark with deionised water.
- Albumin solution
  0.6 g albumin are weighed into a 10 mL volumetric flask and dissolved in deionised water. The flask is then made up to the mark with deionised water.
- Washing solution for the AAS
  100 mL ethanol and 0.5 mL 65% nitric acid are pipetted into a 1000 mL volumetric flask. The flask is made up to the mark with deionised water.

2.4 Calibration standards

- Working solution (1000 µg/L)
  100 µL selenium standard solution (1000 mg selenium/L) and 1 mL 65% nitric acid are pipetted into a 100 mL volumetric flask. The volumetric flask is made up to the mark with deionised water and the solution is homogenised by shaking. The selenium concentration of the working solution is 1000 µg/L.

The working solution should be prepared freshly on a daily basis. Starting with this working solution the calibration standards are prepared according to the pipetting scheme in Table 2, resulting in concentrations up to 240 µg selenium per litre.

Table 2 Pipetting scheme for the preparation of selenium calibration standards.

Volume of working solution (1000 µg/L) [µL]	Final volume of calibration standard [mL]	Selenium concentration [µg/L]
0	10	0
300	10	30
600	10	60
900	10	90
1200	10	120
2400	10	240

To calibrate in matrix, 100 µL calibration standard, 100 µL albumin solution, 600 µL dilution solution and 200 µL matrix modifier solution each are pipetted into a vial.

3 Specimen collection and sample preparation

3.1 Specimen collection

Prior to blood collection, ensure that the blood collection set used is free of selenium. For example, neutral monovettes without addition of coagulation activators are suitable for blood sampling, and lithium-heparine monovettes for plasma sampling. For serum separation the blood sample is allowed to clot for at least 20 min (but no longer than one hour). For plasma separation the blood sample should be mixed thoroughly and centrifuged quickly. After centrifugation (15 min; 2000 × g; 15°C), the supernatant is transferred to neutral sample tubes. Haemolysis must be avoided under all circumstances when collecting serum and plasma samples, respectively.

If selenium determination cannot be performed immediately serum or plasma samples may be kept for a few days in the refrigerator at 4°C. For longer storage e.g. for weeks or months the serum or plasma should be deep frozen at –18°C.

3.2 Sample preparation

For sample preparation 700 µL dilution solution, 200 µL matrix modifier solution and 100 µL serum or plasma sample are pipetted into an autosampler vial.

4 Operational parameters

The operational details given below are intended to be a rough guide only and are especially useful if an identical device type is used. In the case of other types, the parameters are to be optimised accordingly. The setting of additional parameters may become necessary when using spectrometers from other manufacturers.

Wavelength:	196.0 nm
Slit width:	0.5 nm
Injection volume:	20 µL
Injection temperature:	80°C
Measuring time:	3 s
Background correction:	Zeeman
Graphite furnace:	pyrolytic-coated
Evaluation:	extinction

Table 3 Tabular presentation of the temperature program (total time 142 s).

Step	1	2	3	4	5	6	7
Temperature [°C]	100	120	600	600	1450	2100	2400
Increase [°C/s]	2	5	50	0	50	0	0
Time (hold) [s]	20	30	30	5	10	3	3
Gas	Ar	Ar	air	Ar	Ar	Ar	Ar
Gas flow [L/min]	0.2	0.2	0.2	0.2	0.2	off	0.2

5 Analytical determination

20 µL of the 1/10 (v:v) diluted serum or plasma sample (see Section 3) are analysed according to the given opperational parameters using atomic absorption spectrometry. The samples are determined in duplicate and the mean values are used for the calculation of the analytical results.

6 Calibration

The measured extinction of each calibration standard is plotted against the corresponding selenium concentration. The calibration graph obtained is linear in the range between 3 and 240 µg selenium per litre serum or plasma. The calibration graph is given in the Appendix (Figure 1).

Recalibration is necessary if the results of quality control show systematic deviations.

7 Calculation of the analytical result

To determine the analyte concentration in a sample, the extinction measured is inserted into the corresponding calibration graph (see Section 6). The selenium concentration level of the sample is obtained in µg/L. Reagent blank values have to be accounted for where necessary. Normally, calculation of the analytical result is performed by the software of the spectrometer.

8 Standardisation and quality control

Quality control of the analytical results is carried out as stipulated in the guidelines of the *Bundesärztekammer* (German Medical Association) [32].

This means that quality control standards are to be included for precision control within each analytical series. For the quantitation of selenium in serum, control material of different manufacturers and also certified reference material is available. To cover a wide concentration range control material with different concentrations should be used. These should be analysed after calibration, after every twentieth sample and at the end of each analytical series. For external quality control, international quality assessment schemes are available [33, 34].

9 Evaluation of the method

9.1 Precision

To determine the within day precision selenium was analysed ten times in a row using two control materials with nominal values of 81 µg/L and 136 µg/L selenium in serum. The relative standard deviation was determined to be 3.0% and 2.1%, respectively (Table 4). To check the day to day precision selenium was determined in the same control material on twenty different days. The relative standard deviation was determined to be 4.8% (for 81 µg/L Se) and 4.2% (for 136 µg/L Se) (Table 5).

Table 4 Within day precision for the determination of selenium in serum (n = 10).

Analyte	Analyte level [µg/L]	Standard deviation (rel.) s_w [%]	Prognostic range u [%]
Selenium	81	3.0	6.7
	136	2.1	4.7

Table 5 Day to day precision for the determination of selenium in serum (n = 20).

Analyte	Analyte level [µg/L]	Standard deviation (rel.) s_w [%]	Prognostic range u [%]
Selenium	81	4.8	10.0
	136	4.2	8.8

9.2 Accuracy

Accuracy was determined using internal and external quality assurance. For internal quality assurance "Seronorm™ Trace Elements Serum" Level 1 and Level 2 control materials (Sero, Norway) were analysed. The mean values (n = 20) of the day to day precision were 82 and 140 µg/L and are in line with the nominal values of 81 and 136 µg/L.

The quantitation of a spiked serum sample (60 µg/L selenium spiked to a physiological concentration of 40 µg/L) led to a recovery rate of 97%.

For external quality assessment the German External Quality Assessement Scheme (G-EQUAS) No. 39 for Analyses in Biological Materials, carried out by the

Institut und Poliklinik für Arbeits-, Sozial- und Umweltmedizin (Institute and Outpatient Clinic for Occupational, Social and Environmental Medicine) at the University of Erlangen-Nuremberg was participated in [33]. Using the described method selenium concentrations of 79 and 129 µg/L (nominal values: 75 and 133 µg/L) were determined in two plasma samples. Participation in a british external quality assessment scheme [34] also yielded good results. 24, 88 and 157 µg/L were determined in the control materials 859, 860 and 861 from October 2007 (nominal values: 23, 87 and 150 µg/L).

Table 6 Quality assessment for the determination of selenium in serum.

Control material	Analysed value [µg/L]	Nominal value [µg/L]	Recovery rate r [%]
Seronorm™	82	81	101
Trace Elements serum	140	136	103
Spiked serum	97	100	97
G-EQUAS No. 39	79	75	105
	129	133	97
University of Surrey 859, 860 and 861	24	23	104
	88	87	101
	157	150	105

9.3 Detection limit and quantitation limit

A detection limit of 3 µg selenium per litre serum or plasma was estimated for the given procedure. This correspond to a detection limit of 0.3 µg selenium per litre in the measured solution. The detection limit was determined according to the 3 s criterion based on the sensitivity and the standard deviation of the extinction of a blank value solution (n = 10). This corresponds to a quantitation limit of 9 µg/L.

9.4 Sources of error

To avoid interferences by the spectral background the use of Zeeman background correction is required. In addition the temperature program must be carefully optimised in order to minimise background signal during atomisation. To avoid selenium losses prior to the atomisation step a suitable matrix modifier must be selected. The Pd-Mg modifier has proved its value. Interferences from residues in the form of carbon particles in the graphite furnace can be avoided by carrying out ashing with synthetic air. For this reason the ashing procedure is performed in two individual steps (Table 3).

Variations in sensitivity, due to aging of the graphite furnace (e.g. during an extended run), can be compensated for by a dynamic measurement value correction as an additional quality control measure. For this a control standard and a blank value are analysed after every tenth sample and are included in the calculation of the analytical result. This kind of standardisation can be configured in the software of modern spectrometers.

Using graphite furnace AAS no transport interferences are observed. Nevertheless, the pipetting accuracy of the autosampler is affected negatively, when all pipetting and dilution steps are carried out by the autosampler itself (pipetting of the samples, addition of the modifier and of the Triton-X-100 solution). In the case of serum, plasma or blood, a thin film can form on the surface of the autosampler capillary leading to negative effects on pipetting accuracy. Therefore, the author of this method diluted the samples and calibration standards manually to ensure better long-term stability within large analytical series.

The risk of selenium contamination from reagents, equipment or ambient air must be considered at all times. Reagents used must be tested for purity at regular intervals. Rigid purity standards have to be applied for the used equipment, such as vials, volumetric flasks, tubes and pipettes. Clean room conditions are not necessary but useful.

10 Discussion of the method

The method provides accurate and reliable results for selenium levels in serum or plasma using graphite furnace atomic absorption spectrometry with Zeeman background correction. The method is selective, sensitive, easy to handle and therefore suitable for laboratories with a high sample throughput. Due to the high sensitivity of graphite furnace technique, determination is also possible within the concentration range of extreme selenium deficiency in humans. Selenium levels after occupational exposure or intoxication can also be determined by this method. For these determinations the samples must be further diluted if necessary.

For method validation, calibration was carried out using an albumin solution as matrix, because this behaves very similar to serum in the graphite furnace AAS. This procedure was used, as the performance of a standard addition for each individual sample would be a very time consuming process.

As a third calibration strategy one could perform standard additions on the same sample and convert this standard addition calibration into an external calibration for the analysis of the following samples. However, the disadvantage of this calibration procedure is that the selenium levels of a number of real samples would be below the lowest calibration standard.

As selenium also forms a gaseous hydride, hydride-AAS might be used alternatively, ensuring even lower detection limits. However, in a clinical laboratory dealing with a large number of samples, the simple and fast graphite furnace technique should be preferred. Sensitivity, selectivity and robustness are fully sufficient. Finally, compared with the hydride technique, the graphite furnace technique requires less sample material, avoids some additional reagents for digestion and reduction of selenium and is less time-consuming.

Instruments used:

- Thermo Elemental SOLAAR M6 series AA spectrometer with FS95 autosampler and Zeeman background correction (Thermo Elemental)

11 References

1. J. Köhrle: The trace element selenium and the thyroid gland. Biochimie 81, 527–533 (1999).
2. L.N. Vernie: Selenium in carcinogenesis. Biochim. Biophys. Acta 738, 203–217 (1984).
3. K. Schwartz and C.M. Foltz: Selenium as an integral part of factor 3 against dietary necrotic liver degeneration. J. Am. Chem. Soc. 79, 3292–3293 (1957).
4. J.T. Rotruck, A.L. Pope, H.E. Ganther, A.B. Swanson, D.G. Hafeman and W.G. Hoextra: Selenium: biochemical role as a component of glutathione peroxidase. Science 179, 588–590 (1973).
5. B. Michalke: Selenium speciation in human serum of cystic fibrosis patients compared to serum from healthy persons. J. Chromatogr. A 1058, 203–208 (2004).
6. M. Brielmeier and J. Schmidt: Analyse der Funktion von Selenoproteinen. GSF Jahresbericht, GSF, München (2003).
7. D. Behne and A. Kyriakopoulos: Mammalian selenium-containing proteins. Annu. Rev. Nutr. 21, 453–473 (2001).
8. K.H. Goh and T.T. Lim: Geochemistry of inorganic arsenic and selenium in a tropical soil: effect of reaction time, pH, and competitive anions on arsenic and selenium adsorption. Chemosphere 55, 849–859 (2004).
9. A. Kabata-Pendias: Geochemistry of selenium. J. Environ. Pathol. Toxicol. Oncol. 17, 173–177 (1998).
10. Bundesgesetzblatt: Verordnung des Bundesministers für soziale Sicherheit und Generationen über die Qualität von Wasser für den menschlichen Gebrauch (Trinkwasserverordnung – TWV); erste Verordnung zur Änderung der Trinkwasserverordnung vom 03. Mai 2011, BGBl. I Nr. 21 748–774 (2011).
11. H. Greim (ed.): Selen und seine anorganischen Verbindungen. In: Gesundheitsschädliche Arbeitsstoffe, Toxikologisch-arbeitsmedizinische Begründung von MAK-Werten, 29th issue. Wiley-VCH, Weinheim (1999).
12. A. Hartwig (ed.): Selen und seine anorganischen Verbindungen. In: Gesundheitsschädliche Arbeitsstoffe, Toxikologisch-arbeitsmedizinische Begründung von MAK-Werten, 51st issue. Wiley-VCH, Weinheim (2011).
13. L. Thomas: Labor und Diagnose, 4. Aufl., Behring Verlag, 417–419 (1992).
14. I. Lombeck, K. Kasperek, H.D. Harbisch, L.E. Feinendegen and H.J. Bremer: The selenium state of healthy children. I. Serum selenium concentration at different ages; selenium content of food of infants. Eur. J. Pediatr. 125, 81–88 (1973).
15. O. Oster: Zum Selenstatus in der Bundesrepublik Deutschland. Universitätsverlag Jena, 1. Aufl. (1992).
16. J. Piechotowski, U. Weidner, I. Zollner, T. Gabrio, B. Link and M. Schwenk: Serum selenium levels in school children: Results and health assessment. Gesundheitswesen 64, 602–607 (2002).

17 A.C. Muntau, M. Streiter, M. Kappler, W. Röschinger, I. Schmid, A. Rehnert, P. Schramel and A.A. Roscher: Age-related reference values for serum selenium concentrations in infants and children. Clin. Chem. 48, 555–560 (2002).
18 B.E. Jacobson and G. Lockitch: Direct determination of selenium in serum by graphite furnace atomic absorption spectrometry with deuterium background correction and a reduced palladium modifier: Age-specific reference ranges. Clin. Chem. 34, 709–714 (1988).
19 R. Safaralizadeh, G.A. Kardar, Z. Pourpak, M. Moin, A. Zare and S. Teimourian: Serum concentration of selenium in healthy individuals living in Tehran. Nutr. J. 4, 32–35 (2005).
20 H. Al-Sayer, A. Al-Bader. M. Khoursheed, S. Asfar, T. Hussain, A. Behbehani, A. Mathew and H. Dashti: Serum values of copper, zinc and selenium in adults resident in Kuwait. Med. Principles Pract. 9, 139–146 (2000).
21 J. Bleys, A. Navas-Acien and E. Guallar: Serum selenium and diabetes in U.S. adults. Diabetes Care 30, 829–834 (2007).
22 CDC: National report on biochemical indicators of diet and nutrition in the U.S. population 1999–2002. 101–106 (2003).
23 K. Karita, G.S. Hamada and S. Tsugane: Comparison of selenium status between Japanese living in Tokio and Japanese Brazilians in Sao Paulo, Brazil. Asia Pac. J. Clin. Nutr. 10, 197–199 (2001).
24 C.J. Chen, J. S. Lai, C.C. Wu and T.S. Lin: Serum selenium in adult Taiwanese. Sci. Total Environ. 367, 448–450 (2006).
25 R. Raghunath, R.M. Triphathi, S. Mahapatra and S. Sadasivan: Selenium levels in biological matrices in adult population of Mumbai. Sci. Total Environ. 285, 21–27 (2002).
26 K. Hughes, L.H. Chua and C.N. Ong: Serum selenium in the general population of Singapore, 1993 to 1995. Ann. Acad. Med. Singapore 27, 520–523 (1998).
27 J. Burria, M. Haldimann and V. Dudler: Selenium status of the Swiss population: Assessment and change over a decade. J. Trace Elem. Med. Biol. 22, 112–119 (2008).
28 M. Torra, M. Rodamilans, F. Montero and J. Corbella: Serum selenium concentration of healthy northwest Spanish population. Biol. Trace Elem. Res. 58, 127–133 (1997).
29 C.D. Thomson: Assessment of requirements for selenium and adequacy of selenium status: a review. Eur. J. Clin. Nutr. 58, 391–402 (2004).
30 A. Hartwig and H. Drexler (eds.): Selen und seine anorganischen Verbindungen. Biologische Arbeitsstoff-Toleranz-Werte (BAT-Werte) und Expositionsäquivalente für krebserzeugende Arbeitsstoffe (EKA) und Biologische Leitwerte (BLW), 18th issue, Wiley-VCH, Weinheim (2011).
31 Robert Koch-Institut: Selen in der Umweltmedizin. Bundesgesundheitsbl. Gesundheitsforsch. Gesundheitsschutz 49, 88–102 (2006).
32 Bundesärztekammer: Richtlinie der Bundesärztekammer zur Qualitätssicherung quantitativer laboratoriumsmedizinischer Untersuchungen. Dt. Ärztebl. 100, A3335–A3339 (2003).
33 The German External Quality Assessment Scheme (G-EQUAS) No. 39 for Analyses in Biological Materials. Institute and Outpatient Clinic of Occupational, Social and Environmental Medicine of the University of Erlangen-Nuremberg, Germany (2007).
34 Trace elements external quality assessment scheme, University of Surrey, Guildford, Surrey, UK (2007).

Author: *P. Heitland*
Examiner: *B. Michalke*

12 Appendix

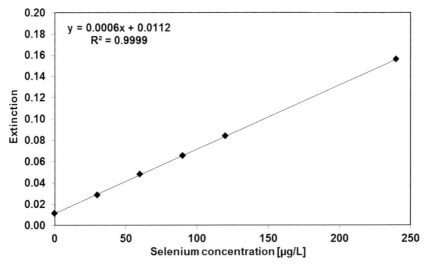

Fig. 1 Calibration graph for the determination of selenium in serum.

Tetrahydrofuran (THF) in urine (Addendum to the DFG method "Alcohols and Ketones")

Matrix:	Urine
Hazardous substance:	Tetrahydrofuran
Analytical principle:	Headspace gas chromatography/flame ionisation detection (headspace GC/FID)
Completed in:	March 1998

Overview of parameters that can be determined with this method and the corresponding chemical substances:

Hazardous substance	CAS	Parameter	CAS
Tetrahydrofuran	109-99-9	Tetrahydrofuran	109-99-9

Summary

This method permits sensitive and specific quantitation of tetrahydrofuran (THF) in the urine of persons occupationally exposed to this substance. The determination of THF in urine is based on the DFG method "Alcohols and Ketones" from February 1996 [1].

The volatile THF contained in the urine is determined using headspace capillary gas chromatography. For this purpose, the urine samples are heated in gas-tight sealed crimp cap vials to 40°C. When partition equilibrium of the tetrahydrofuran between liquid and gas phase is reached a headspace gas phase aliquot is analysed by gas chromatography. As detector, a flame ionisation detector (FID) is used.

Analyte concentrations are calculated using standard calibration graphs. Calibration standard solutions are prepared by spiking pooled human urine with the standard substance. These calibration standard solutions are treated in the same manner as the urine samples to be investigated.

Tetrahydrofuran

Within day precision:	Standard deviation (rel.)	$s_w = 2.15\%$
	Prognostic range	$u = 4.84\%$
	at a spiked concentration of 3.55 mg tetrahydrofuran per litre urine and where n = 10 determinations	
Day to day precision:	Standard deviation (rel.)	$s_w = 2.44\%$
	Prognostic range	$u = 6.25\%$
	at a spiked concentration of 3.55 mg tetrahydrofuran per litre urine and where n = 6 determinations	
Accuracy:	Recovery rate	r = 99.5–101.3%
	at a nominal concentration of 3.55 mg tetrahydrofuran per litre urine and where n = 6–10 determinations	
Detection limit:	0.1 mg tetrahydrofuran per litre urine	
Quantitation limit:	0.3 mg tetrahydrofuran per litre urine	

Tetrahydrofuran (THF)

Tetrahydrofuran is a colourless solvent with an acetone-like odour. The odour threshold is about 30 mL/m^3 [1]. THF is readily miscible with water and most organic solvents. It has many uses, e.g. in glues, paints, lacquers and detergents. One important application is the production of magnetic tapes. At the workplace, it is mainly taken up by inhalation. In most cases mixed exposures are involved, as THF frequently occurs together with other organic solvents. THF can also be absorbed through the skin.

In spite of the widespread use in industry no adverse health effects after inhalation or skin absorption of THF have been described up to now. The odour of THF, which is initially perceived to be pleasant, produces impairments of well-being after a short time. This warning signal generally prevents exposure to higher concentrations when working with THF.

Studies on THF metabolism in humans are not available. At least part of the absorbed THF is excreted unchanged with the urine [2, 3].

Investigations on non-exposed persons have shown that, without occupational exposure, no THF is found in the urine up to a detection limit of 100 µg/L. The urinary excretion of unmetabolised THF correlates significantly (r = 0.86) with the external THF concentration. The corresponding correlation factor for THF in blood is clearly worse with r = 0.68 [3]. Therefore, a BAT value (*Biologischer Arbeitsstoff-Toleranzwert*, biological tolerance value) of 2 mg THF per litre urine was established [4, 5].

Author: *M. Blaszkewicz*
Examiner: *J. Angerer*

Tetrahydrofuran (THF) in urine (Addendum to the DFG method "Alcohols and Ketones")

Matrix:	Urine
Hazardous substance:	Tetrahydrofuran
Analytical principle:	Headspace gas chromatography/flame ionisation detection (headspace GC/FID)
Completed in:	March 1998

Contents

1	General principles
2	Equipment, chemicals and solutions
2.1	Equipment
2.2	Chemicals
2.3	Calibration standards
3	Specimen collection and sample preparation
4	Operational parameters
4.1	Operational parameters for gas chromatography
5	Analytical determination
6	Calibration
7	Calculation of the analytical result
8	Standardisation and quality control
9	Evaluation of the method
9.1	Precision
9.2	Accuracy
9.3	Detection limits and quantitation limits
9.4	Sources of error
10	Discussion of the method
11	References
12	Appendix

1 General principles

The procedure described here permits the sensitive and specific quantification of tetrahydrofuran in the urine of persons occupationally exposed to this substance. The determination of THF in urine is based on the DFG-method "Alcohols and Ketones" of February 1996 [1].

The volatile THF contained in the urine is determined using headspace gas chromatography. For this purpose, the urine samples are heated to 40°C in gas-tight sealed crimp cap vials. When partition equilibrium of the tetrahydrofuran between liquid and gas phase is reached a headspace gas phase aliquot is analysed by gas chromatography (GC). A flame ionisation detector (FID) is used for detection purposes.

Analyte concentrations are calculated using standard calibration graphs. Calibration standard solutions are prepared by spiking pooled human urine with the standard substance. These calibration standard solutions are treated in the same manner as the urine samples to be investigated.

2 Equipment, chemicals and solutions

2.1 Equipment

- Gas chromatograph with flame ionisation detector (FID), automatic headspace sampler and data processing system for evaluation.
- Gas chromatographic column: length: 10 m; particle trap: 2.5 m; inner diameter: 0.32 mm; stationary phase: vinyl benzene divinylbenzene polymer; film thickness: 10 μm (e.g. CP PoraPLOT Q, Agilent Technologies Germany GmbH)
- 250 mL Urine containers with screw caps (e.g. Sarstedt No. 77.577)
- 20 mL Crimp cap vials with butyl rubber/PTFE (polytetrafluoroethylene) septa and aluminium crimp caps (e.g. Macherey-Nagel N 20-20 HS
- Pneumatic crimping tool (e.g. Macherey-Nagel)
- 10 mL and 25 mL volumetric flasks (e.g. Brand)
- Piston-stroke pipettes with adjustable volume between 10–100 μL or 100–1000 μL with suitable pipette tips (e.g. Eppendorf)
- Analytical balance (e.g. Sartorius)

2.2 Chemicals

- Tetrahydrofuran p.a. (e.g. Sigma-Aldrich, No. 34865)
- Bidist. water
- Nitrogen 5.0 (e.g. Linde)
- Hydrogen 5.0 (e.g. Linde)
- Synthetic air, hydrocarbon free (e.g. Linde)

2.3 Calibration standards

- Stock solution (49.31 mM THF; 3.56 g THF/L)
  100 µL THF are pipetted into a 25 mL volumetric flask. The flask is then made up to the mark with bidist. water.
- Working solution (1.97 mM THF; 0.142 g THF/L)
  1 mL stock solution is pipetted into a 25 mL volumetric flask. The flask is then made up to the mark with bidist. water.

Calibration standards

Starting with the working solution, calibration standards are prepared in urine by pipetting between 30 µL and 500 µL of the working solution into 10 mL volumetric flasks and making up to the mark with pooled urine (Table 1). This results in concentrations between 0.426 mg and 7.1 mg THF per litre urine. The calibration standards are freshly prepared each day.

Table 1 Pipetting scheme for the preparation of the THF calibration standards in urine.

Volume of the working solution [mL]	Final volume of the calibration standard solution [mL]	Analyte level [mg/L]
0.5	10	7.10
0.25	10	3.55
0.1	10	1.42
0.05	10	0.71
0.030	10	0.426

In each case, 2 mL of the calibration standard are pipetted into crimp cap vials. The vials are immediately sealed with butyl rubber/PTFE-septa and aluminium crimp caps.

3 Specimen collection and sample preparation

Specimens should be collected at the end of a shift, or better at the end of a working week or after several previous shifts [5]. The urine samples are collected in sealable plastic containers. In each case, 2 mL of each urine sample is pipetted immediately after specimen collection into crimp cap vials kept ready for this purpose. The samples should be transported under cooled conditions and should be kept in a refrigerator up to analysis. If storage over a longer period is necessary, the samples can be stored in the deep freezer at –18°C up to analysis. Before analysis the samples are brought to room temperature. The crimp cap vials are placed in the heating block of the headspace autosampler and incubated for one hour at

40 °C (incubation time: 45 min; mixing: 10 min; stabilisation: 5 min). This ensures the adjustment of the vapour-liquid partition equilibrium.

4 Operational parameters

4.1 Operational parameters for gas chromatography

Capillary column:	Material:	Fused Silica PoraPlot Q (e.g. Agilent)
	Stationary phase:	PoraPak Q
	Length:	10 m and 2.5 m particle trap
	Inner diameter:	0.32 mm
	Film thickness:	10.0 µm
Detector:	Flame ionisation detector (FID)	
Temperature:	Column:	Initial temperature 100 °C; 4 min isothermal, then increase at a rate of 10 °C/min to 140 °C; 2 min isothermal; then increase at 5 °C/min to 200 °C; 1 min isothermal
	Injector:	220 °C
	Detector:	250 °C
Carrier gas:	Nitrogen 5.0 at a column pre-pressure of 40 kPa	
Split:	splitless	
Injection volume:	1000 µL	

All other parameters must be optimised in accordance with the manufacturer's instructions.

5 Analytical determination

After incubation of the crimp cap vials for one hour at 40 °C, 1000 µL of the headspace phase is injected into the GC/FID system.

With each series at least one urine control sample and one reagent blank are analysed. For the latter, bidist. water is pipetted into the crimp cap vials instead of urine.

Figure 1 (in the Appendix) shows as an example the chromatogram of a spiked urine sample. The THF retention time of 9.4 min shown in Figure 1 is intended to be a rough guide only. Users of the method must ensure proper separation performance of the capillary column they use and of the resulting retention behaviour of the analyte.

6 Calibration

The calibration standard solutions in urine prepared according to Section 2.3 are incubated according to Section 3 and analysed by headspace GC/FID. An unspiked urine sample is used for the determination of the blank value. The calibration graph is obtained by plotting the THF peak areas as a function of the spiked analyte levels. The unspiked urine blank is subtracted to compensate for the background. It is not necessary to plot a complete calibration graph for every analytical series. As a rule, it is sufficient to include one calibration standard in every analytical series. In that case, the peak area of the one calibration standard is calculated and compared to the peak area of the equivalent standard in the complete calibration function. A new calibration graph should be prepared if the quality control results indicate a systematic deviation. The calibration function for THF is linear between 0.071 mg and 56.8 mg per litre urine. Figure 2 (in the Appendix) shows an example of a calibration graph of THF in urine.

7 Calculation of the analytical result

The analyte concentration in a sample is calculated on the basis of the calibration graph according to Section 6. With the peak area obtained for the analyte the corresponding THF concentration in mg/L can be calculated via the calibration function. A possible reagent blank value has to be taken into consideration.

8 Standardisation and quality control

Quality control of the analytical results is carried out as stipulated in the guidelines of the *Bundesärztekammer* (German Medical Association) [6]. For quality control, urine spiked with a defined amount of THF is analysed within each analytical series. A quality control material for THF in urine is not commercially available. Thus, it has to be prepared in the laboratory. For this purpose the urine of individuals occupationally non-exposed to THF is spiked with a defined quantity of THF in the relevant concentration range. The control urine is aliquoted into headspace vials. Stored in a deep-freezer at about –20°C, it can be used for up to six months. The nominal value and the tolerance ranges of this quality control material are determined in a pre-analytical period (one analysis of the control material on each of 15 different days) [6–8].

9 Evaluation of the method

9.1 Precision

To determine precision, defined quantities of THF are spiked to pooled urine of persons, occupationally non-exposed to THF, to obtain a concentration of 3.55 mg/L. For within day precision the following relative standard deviations with the corresponding prognostic ranges were obtained:

Table 2 Within day precision for the determination of tetrahydrofuran in urine (n = 9–10).

Analyte	Spiked concentration [mg/L]	Standard deviation (rel.) s_w [%]	Prognostic range u [%]
Tetrahydrofuran Series 1 (n = 10)	3.55	2.15	4.86
Tetrahydrofuran Series 2 (n = 9)	3.55	1.73	3.99

Furthermore, the precision from day to day was determined. For this purpose the spiked urine was processed on six different days and analysed as described in the preceding sections. The obtained precision data are shown in Table 3.

Table 3 Precision from day to day for the determination of tetrahydrofuran in urine (n = 6).

Analyte	Spiked concentration [mg/L]	Standard deviation (rel.) s_w [%]	Prognostic range u [%]
Tetrahydrofuran	3.55	2.44	6.25

9.2 Accuracy

The accuracy of the method was determined by spiking the pooled urine of persons occupationally non-exposed to THF with 3.55 mg THF per litre urine and

Table 4 Recovery rates for tetrahydrofuran in urine (n = 6–10).

Analyte	Concentration after spiking [mg/L]	Recovery rate r [%]	Range [%]
Tetrahydrofuran Series 1 (n = 10)	3.55	101.3	97.6–103.7
Tetrahydrofuran Series 2 (n = 9)	3.55	99.5	97.4–102.1
Tetrahydrofuran Series 3 (n = 6)	3.55	100.9	97.5–105.2

dividing it into aliquots. To determine recovery rates the urine was processed and analysed six to ten times as described in Section 3. This yielded mean recovery rates between 99.5% and 101.3%. Analyte losses due to sample preparation were not determined.

9.3 Detection limits and quantitation limits

The detection limit was 0.1 mg per litre urine (three times the signal/background noise ratio) under the conditions given for sample preparation and determination by means of GC/FID. This corresponds to a quantitation limit of 0.3 mg THF per litre urine.

9.4 Sources of error

The sampling time prescribed in the List of MAK and BAT Values [9] must be adhered to under all circumstances. Specimens should be collected at the end of a shift, or better at the end of a working week or several previous shifts [5, 9].

In order to avoid loss of the volatile THF during transport and storage, care must be taken that the crimp cap vials are sealed gas-tight. It should not be possible to turn the cap by hand any more. Make sure the crimping tongs are adjusted correspondingly.

Furthermore, care must be taken that both the crimp cap vials as well as the butyl rubber/PTFE septa have been baked out before use. Bake-out should be done at 100°C for three days in the drying oven.

For precision and accuracy of analytical results it is of decisive importance that the vapour-liquid partition equilibrium is reached. Experience has shown that, in the case of THF, the equilibrium has been reached after one hour. In spite of this, the user of this method should check when, with the apparatus used and under the prevailing conditions, the partition equilibrium has been reached.

10 Discussion of the method

With the Addendum to "Alcohols and Ketones in Blood and Urine" presented here, it is possible to determine THF in addition to eleven further alcohols and ketones in urine. The analytical procedure given represents a suitable and validated method for the determination of THF in the urine of occupationally exposed persons.

The method is simple to perform, as in the case of headspace GC it is possible to do without sample preparation. An analytical background does not occur, so that a very low detection limit of 100 µg THF per litre urine can be obtained. The reliability criteria of the method are described as good. The sensitivity can be considered sufficient for the lower concentration range as well.

With the 10 m PoraPLOT-Q-column used, the author of the method identified a co-elution of 2-butanone and THF so that 2-butanone must be absent when using this column. The examiner of this method was, however, able to obtain a good separation of these two substances using a longer column (HP-Plot 30 m × 0.32 mm × 20 µm).

The determination of THF in blood is also possible analogous to the DFG method "Alcohols and Ketones" [1]. The calibration function of THF in blood is strictly linear over a concentration range of 0.89 to 8.89 mg THF/L blood. The within day precision indicates a relative standard deviation of 3.42% at a spiked concentration of 1.78 mg THF per litre blood and n = 9 determinations. In the scope of this Addendum, it was decided not to undertake any further validation of the THF determination in blood, as the THF concentration in blood correlates less accurately with the external THF concentration (r = 0.68) than the THF concentration in urine (r = 0.86) [3].

Instruments used:

- Gas chromatograph Shimadzu GC 14A PSF with FID and injector for packed columns and Wide Bore Column Attachment, headspace Tekmar 7000 autosampler with Carousel 7050 and Septum Needle Adaptor, Integrator HP 3394A.

11 References

1 H. Greim, J. Angerer, K.H. Schaller (eds.): Alkohols and Ketones in blood and urine. Analyses of Hazardous Substances in Biological Materials. Volume 5, VCH, Weinheim (1997).
2 H. Greim and G. Lehnert (eds.): Tetrahydrofuran. BAT Value Documentations. Volume 2, VCH, Weinheim (1999).
3 C.N. Ong, S.E. Chia, W.H. Phoon, K.T. Tan: Biological monitoring of occupational exposure to tetrahydrofuran. Br. J. Ind. Med. 48, 616–621 (1991).
4 H. Greim (ed.): Tetrahydrofuran. Gesundheitsschädliche Arbeitsstoffe, Toxikologisch-arbeitsmedizinische Begründung von MAK-Werten, 37th issue, Wiley-VCH, Weinheim (2003).
5 H. Greim and H. Drexler (eds.): Tetrahydrofuran, Addendum. The MAK-Collection Part II: BAT Value Documentations. Volume 4, Wiley-VCH, Weinheim (2005).
6 Bundesärztekammer: Richtlinie der Bundesärztekammer zur Qualitätssicherung quantitativer laboratoriumsmedizinischer Untersuchungen. Dt. Ärztebl. 105, A341–A355 (2008).
7 J. Angerer and G. Lehnert: Anforderungen an arbeitsmedizinisch-toxikologische Analysen – Stand der Technik. Dt. Ärztebl 37, C1753–C1760 (1997).
8 J. Angerer, T. Göen, G. Lehnert: Mindestanforderungen an die Qualität von umweltmedizinisch-toxikologischen Analysen. Umweltmed. Forsch. Prax. 3, 307–312 (1998).

9 Deutsche Forschungsgemeinschaft (DFG): List of MAK and BAT Values 2012: Maximum Concentrations and Biological Tolerance Values at the Workplace. Report No. 48. Wiley-VCH, Weinheim (2012).

Author: *M. Blaszkewicz*
Examiner: *J. Angerer*

12 Appendix

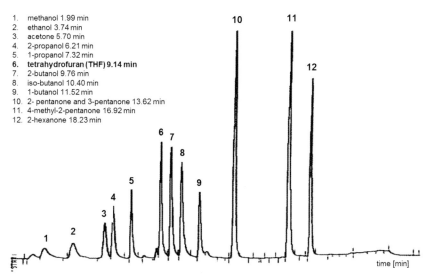

Fig. 1 Chromatogram of a urine standard: separation of THF from eleven further alcohols and ketones.

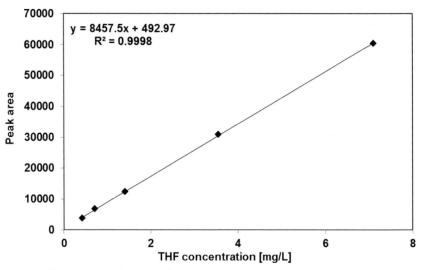

Fig. 2 Calibration graph of tetrahydrofuran in urine.

Thiocyanate in plasma and saliva

Matrix:	Plasma and saliva
Hazardous substances:	Hydrogen cyanide, cyanides and cyanide releasing chemicals
Analytical principle:	Photometry in microtiter plates
Completed in:	October 1998

Overview of the parameters that can be determined with this method and the corresponding hazardous substances:

Hazardous substance	CAS	Parameter	CAS
Hydrogen cyanide	74-90-8		
Cyanides	57-12-5		
Sodium cyanide	143-33-9		
Potassium cyanide	151-50-8	Thiocyanate	302-04-5
Cyanogen chloride	506-77-4		
Oxalic acid dinitrile	460-19-5		
Acetonitrile	75-05-8		
Acrylonitrile	107-13-1		

Summary

Thiocyanate (rhodanide) is the main metabolite of cyanide and can thus be used as biomarker for exposure to cyanide or to cyanide releasing chemicals. Especially for chronic exposure to low cyanide concentrations, such as occur for example in smoking and at a number of workplaces, thiocyanate in plasma and saliva is a suitable biomarker. The procedure described here is based on a method published by Degiampietro et al. [1] and permits the rapid and reliable determination of thiocyanate in plasma and saliva. It is a photometric method performed in 96-well plates for high throughput using a plate reader. When Fe(III) ions are added to samples containing thiocyanate (SCN^-), a red complex is formed, which is measured close to its absorption maximum at 492 nm. After addition of mercury(II) nitrate, which forms a colourless $[Hg(SCN)_4]^{2-}$-complex, the sample blank value is determined and subtracted.

Thiocyanate in plasma

Within day precision:	Standard deviation (rel.)	s_w = 17.4% or 4.2%
	Prognostic range	u = 48.4% or 10.8%
	at a concentration of 1.33 mg or 12.4 mg thiocyanate per litre plasma and where n = 5 or 6 determinations	
Day to day precision:	Standard deviation (rel.)	s_w = 12.8% or 5.1%
	Prognostic range	u = 30.2% or 12.0%
	at a concentration of 2.49 mg or 11.0 mg thiocyanate per litre plasma and where n = 8 determinations	
Accuracy:	Recovery rate (rel.)	r = 114.8% or 98.4%
	at a nominal concentration of 2.71 mg or 12.1 mg thiocyanate per litre plasma and where n = 4 determinations	
Detection limit:	0.76 mg thiocyanate per litre plasma	
Quantitation limit:	2.28 mg thiocyanate per litre plasma	

Thiocyanata in saliva

Within day precision:	Standard deviation (rel.)	s_w = 1.4% or 2.9%
	Prognostic range	u = 3.9% or 8.1%
	at a concentration of 38 mg or 167 mg thiocyanate per litre saliva and where n = 5 determinations	
Day to day precision:	Standard deviation (rel.)	s_w = 2.7% or 1.2%
	Prognostic range	u = 6.4% or 2.8%
	at a concentration of 36 mg or 162 mg thiocyanate per litre saliva and where n = 8 determinations	
Accuracy:	Recovery rate (rel.)	r = 103% or 100%
	at a nominal concentration of 38.9 mg or 112.7 mg thiocyanate per litre saliva and where n = 4 determinations	
Detection limit:	0.76 mg thiocyanate per litre saliva	
Quantitation limit:	2.28 mg thiocyanate per litre saliva	

Thiocyanate

Detoxification of cyanide, a potent inhibitor of cellular respiration, occurs primarily through formation of thiocyanate. The major mechanism to form thiocyanate in the human body is the enzymatic transfer of a sulfur atom from thiosulfate to the cyanide by thiosulfate sulfurtransferase (rhodanase) [2] (Figure 1). Thiocyanate in body fluids (plasma, saliva, urine) can therefore be used as a biomarker of exposure to cyanides or to cyanide-releasing chemicals. On account of the relatively long half-life of 6–14 days in the mentioned body fluids [3, 4], thiocyanate is especially suitable for the detection of chronic low-dose exposure to cyanide, whereas the determination of cyanide in blood is mainly used for acute cyanide exposure [2].

Gaseous hydrogen cyanide is liberated from cyanide salts when in contact with acids or carbon dioxide and is thus occurring wherever cyanides are handled. This

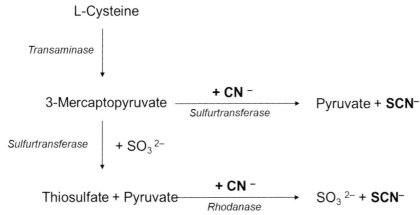

Fig. 1 Thiocyanate formation from cyanide acc. to [5]. Thiocyanate is formed by enzymatic transfer of sulfur from 3-mercaptopyruvate (via sulfurtransferase) or from thiosulfate (via rhodanase) to cyanide. The formation of thiocyanate mainly takes place via rhodanase.

is the case among others where galvanic baths are used. Hydrogen cyanide is in addition readily formed from nitriles as acetonitrile, acrylonitrile and cyanohydrins [6, 7] and is released from the combustion of nitrogen-containing plastics. In addition, hydrogen cyanide is used as a fumigant on ships.

Compared with non-smokers, smokers show 2 to 3 times higher thiocyanate levels (Table 1). The mainstream smoke of a cigarette contains about 50–200 µg hydrogen cyanide [8]. Up to the 1980s, thiocyanate levels were frequently used to differentiate between smokers and non-smokers, and as an objective measure for exposure to tobacco smoke. Today, cotinine is used as a more specific biomarker of tobacco smoke exposure [9]. The specificity of thiocyanate as a marker of exposure to low cyanide concentrations at the workplace or from active smoking is limited by the fact that cyanides or thiocyanate occur also in several foods. Cyanogenic glycosides occur in almonds, nuts, pulses, bamboo shoots, beans, linseed and beer.

Table 1 Thiocyanate levels in plasma and saliva of non-smokers and smokers (MV ± SD).

Non-smokers	Smokers	Reference
Thiocyanate in plasma (mg/L)		
3.47 ± 2.39 (n = 6815)	9.09 ± 3.41 (n = 10377)	Bliss and O'Connell, 1984 [11]
3.16 ± 1.75 (n = 1356)	10.10 ± 3.22 (n = 5090)	Ockene et al., 1987 [12]
3.08 ± 1.59 (n = 3274)	10.04 ± 3.03 (n = 4553)	Ruth and Neaton, 1991 [13]
Thiocyanate in saliva (mg/L)		
70.9 ± 44.2 (n = 242)	158 ± 64.5 (n = 287)	Bliss and O'Connell, 1984 [11]
75.5 (n = 100)	142 (n = 94)	Jarvis et al., 1984 [14]
97.0 (median) (n = 207)	170 (median) (n = 117)	Degiampietro et al., 1987 [1]

Furthermore cyanides are present in the seeds of pome and stone fruits and they are in that way also present in fruit brandies. Preformed thiocyanate (in form of glucosinolates) occurs in cabbage, root vegetables, mustard and milk [10, 11]. These sources make it generally difficult to evaluate thiocyanate levels as biomarker of exposure for cyanide.

Authors: *K. Riedel, H. W. Hagedorn, G. Scherer*
Examiner: *J. Angerer*

Thiocyanate in plasma and saliva

Matrix:	Plasma and saliva
Hazardous substances:	Hydrogen cyanide, cyanides and cyanide releasing chemicals
Analytical principle:	Photometry in microtiter plates
Completed in:	October 1998

Contents

1	General principles
2	Equipment, chemicals and solutions
2.1	Equipment
2.2	Chemicals
2.3	Solutions
2.4	Calibration standards
3	Specimen collection and sample preparation
3.1	Specimen collection
3.2	Sample preparation
4	Operational parameters
5	Analytical determination
5.1	Pipetting scheme for the microtiter plate
5.2	Test procedure
5.3	Photometric determination
6	Calibration
7	Calculation of the analytical results
8	Standardisation and quality control
9	Evaluation of the method
9.1	Precision
9.2	Accuracy
9.3	Detection limits and quantitation limits
9.4	Sources of error
10	Discussion of the method
11	References
12	Appendix

1 General principles

Thiocyanate (SCN⁻) in plasma and saliva is determined photometrically without previous deproteinisation or hydrolysis. With SCN⁻, Fe(III) ions form the red Fe[SCN]$_3$ complex with an absorption maximum at 492 nm. Chloride also forms a coloured complex with Fe(III) ions. Spectral overlap, however, is negligible at wavelengths above 460 nm. A reagent blank value for all samples as well as a blank value for each saliva or plasma sample is taken into account. The sample blank value is determined after addition of mercury(II) nitrate, which forms a colourless [Hg(SCN)$_4$]$^{2-}$ complex.

2 Equipment, chemicals and solutions

2.1 Equipment

- Multiplate reader with a 492 nm filter (e.g. Tecan GENios™ with plate shaking function and Software Magellan 4.0)
- Microcentrifuge (e.g. EBA 12R, Hettich)
- Centrifuge for blood samples (e.g. Rotanta 460R, Hettich)
- Stroke pipettes with adjustable volume between 10–100 µL or 100–1000 µL with suitable pipette tips (e.g. Eppendorf)
- Multipette™ Plus with Combitips for 25 µL (e.g. Eppendorf)
- Transparent 96-well microtiter plate for absorption measurements with lid (e.g. Costar 96-Well Cell Culture Cluster)
- 10 mL, 100 mL, 1000 mL Volumetric flasks (e.g. Brand)
- Potassium EDTA-S monovettes™ for blood sampling (e.g. Sarstedt)
- Salivettes™ for saliva sampling (e.g. Sarstedt)
- Aluminium foil

2.2 Chemicals

- Fe(III) nitrate nonahydrate p.a. (e.g. Merck, No. 103883)
- Mercury(II) nitrate monohydrate p.a. (e.g. Merck, No. 104439)
- Nitric acid 65% p.a. (e.g. Merck, No. 100456)
- Polyoxyethylene laurylether (Brij™ 35) (e.g. Merck, No. 801962)
- Ammonium thiocyanate p.a. (e.g. Merck, No. 101213)
- Bidistilled water

2.3 Solutions

- 150 mM nitric acid
  300 mL bidist. water are placed in a 1000 mL volumetric flask, and 10.4 mL 65% nitric acid are added. The flask is then made up to the mark with bidist. water.
- 270 mM nitric acid
  300 mL bidist. water are placed in a 1000 mL volumetric flask, and 18.7 mL 65% nitric acid are added. The flask is then made up to the mark with bidist. water.
- Reaction solution A (40 mM Fe(III) nitrate nonahydrate)
  1.62 mg Fe(III) nitrate nonahydrate and 6 g Brij™ 35 are weighed into a 100 mL volumetric flask and dissolved in about 70 mL 270 mM nitric acid. The flask is then made up to the mark with 270 mM nitric acid.
- Reaction solution B (60 mM mercury(II) nitrate monohydrate)
  2.06 mg mercury(II) nitrate monohydrate are weighed into a 100 mL volumetric flask, and the flask is then made up to the mark with 150 mM nitric acid.

2.4 Calibration standards

- Stock solution I (6.57 mM ammonium thiocyanate)
  50 mg ammonium thiocyanate are weighed into a 100 mL volumetric flask and the flask is made up to the mark with bidist. water. The stock solution should be kept in the refrigerator at a temperature of +4°C. Under these conditions it is stable for at least 10 weeks.
- Stock solution II (0.657 mM ammonium thiocyanate)
  1 mL stock solution I is placed in a 10 mL volumetric flask, and the flask is then made up to the mark with bidist. water. The stock solution should be kept in the refrigerator at a temperature of +4°C. Under these conditions it is stable for at least 10 weeks.
- Calibration standards
  The ammonium thiocyanate stock solution I (500 mg/L) is diluted with bidist. water according to the pipetting scheme given in Table 2, the ammonium thiocyanate stock solution II (50 mg/L) according to the pipetting scheme given in Table 3. The calibration standards are stored in sealed volumetric flasks or 10 mL polyethylene vials. The prepared calibration standards are storable for at least 10 weeks at a temperature of +4°C.

Table 2 Determination of thiocyanate in saliva: pipetting scheme for the preparation of calibration standards in water.

Volume of stock solution I (500 mg/L) [mL]	Final volume of calibration standard solution [mL]	Concentration level ammonium thiocyanate [mg/L]	Analyte level thiocyanate [mg/L]
1	1	500	382
1	2	250	191
1	5	100	76.3
0.5	5	50	38.2
0.25	5	25	19.1
0.25	10	12.5	9.54

Table 3 Determination of thiocyanate in plasma: pipetting scheme for the preparation of calibration standards in water.

Volume of stock solution II (50 mg/L) [mL]	Final volume of calibration standard solution [mL]	Concentration level ammonium thiocyanate [mg/L]	Analyte level thiocyanate [mg/L]
1	1	50.0	38.2
1	2	25.0	19.1
1	5	10.0	7.63
0.5	5	5.0	3.82
0.25	5	2.5	1.91
0.25	10	1.25	0.954

3 Specimen collection and sample preparation

3.1 Specimen collection

Blood samples are collected with an EDTA monovette. The fresh blood samples are mixed thoroughly and allowed to equilibrate to room temperature. Then, they are centrifuged immediately for 10 min at $3000 \times g$ in a refrigerated centrifuge (+10°C) to separate the cellular components. The supernatant plasma is pipetted off. For thiocyanate determination fresh plasma or plasma which has been frozen at −20°C for only a few days should be used.

Saliva collection is carried out by chewing a cotton wool swab (e.g. Salivette[TM], Sarstedt). The saliva sample obtained from the swab is centrifuged at $3000 \times g$ for 10 min. The supernatant is pipetted off and can be stored at −20°C until further processing.

3.2 Sample preparation

Deep frozen samples are thawed at room temperature. The plasma or saliva is centrifuged in the microcentrifuge at 3220 × g (+10°C) for 10 min.

4 Operational parameters

Photometric determination of the sample is performed in 96-well microtiter plates using a Tecan GENios™ multiplate reader.

Operational parameters for UV spectroscopy:

Measurement technique:		Absorbance
Measurement wavelength:		492 nm
Measurement direction:		in columns
Number of determinations:		Three per sample
Plate shaking:	Duration:	180 s
	Mode:	orbital
	speed:	low
Reset phase:		300 s
Temperature:		36°C ± 0.5°C

5 Analytical determination

5.1 Pipetting scheme for the microtiter plate

The samples and the reagent blank values are pipetted into four wells each. The calibration standards S1 to S6 are pipetted into two wells each. Column 1 and 2 are used for the blank values (BV) and the standards (S1 to S6), the remaining columns for the samples P1 to P20 (see Figure 2).

5.2 Test procedure

According to the pipetting scheme for the microtiter plate, 25 µL bidist. water (as reagent blank), 25 µL standard or 25 µL sample are pipetted into the well. Then, 75 µL bidist. water and 125 µL reaction solution A are pipetted into each well. The 96 well plate is shaken on the orbital shaker for 10 min. During this process, the plate should be protected from light with aluminium foil.

BV	BV	P1	P3	P5	P7	P9	P11	P13	P15	P17	P19
S6	S6	P1	P3	P5	P7	P9	P11	P13	P15	P17	P19
S5	S5	P1	P3	P5	P7	P9	P11	P13	P15	P17	P19
S4	S4	P1	P3	P5	P7	P9	P11	P13	P15	P17	P19
S3	S3	P2	P4	P6	P8	P10	P12	P14	P16	P18	P20
S2	S2	P2	P4	P6	P8	P10	P12	P14	P16	P18	P20
S1	S1	P2	P4	P6	P8	P10	P12	P14	P16	P18	P20
BV	BV	P2	P4	P6	P8	P10	P12	P14	P16	P18	P20

Fig. 2 Pipetting scheme for the microtiter plate.

5.3 Photometric determination

The measurement on the Tecan GENios™ multiplate reader is performed computer-assisted at 492 nm with reference at 620 nm (using the Magellan 4.0 software).

After starting, the plate is measured automatically. After the first measurement, the plate must be removed from the reader and 25 µL reaction solution B are pipetted into each well within 4 min. The plate is then placed again in the reader and automatically measured for a second time after 5 min. The extinction difference from both measurements is then calculated for all standards and samples.

6 Calibration

For evaluation, the measured extinctions must be exported into a spreadsheet program (e.g. Microsoft Excel) and processed accordingly. The blank adjusted extinction differences of the respective calibration standards are averaged and plotted against the corresponding thiocyanate concentrations. A linear relationship between absorption and concentration is obtained for both concentration ranges (determination of thiocyanate in plasma or saliva). The calibration graphs are given in the Appendix (Figures 3 and 4).

7 Calculation of the analytical results

To determine the analyte concentration in a sample, the mean extinction difference of the blank value is subtracted from the mean extinction difference of the sample. The value thus obtained is inserted into the corresponding calibration graph (see Section 6). The SCN⁻ concentration level of the sample is obtained in mg/L.

8 Standardisation and quality control

Statistical quality control is carried out using control samples prepared in the laboratory. Control material, prepared from a saliva pool derived from various donors, is divided into 150 µL aliquots and stored at –20°C. One of these samples is analysed within each analytical series. Thus, information on the precision of the method and on the long-term stability of the samples is obtained. The use of a plasma pool is not recommended, as precipitates may be formed in the plasma after long-term storage, which affects the photometric determination adversely.

9 Evaluation of the method

9.1 Precision

To determine the within day precision, four saliva samples and one plasma sample spiked with three different thiocyanate levels were analysed at least five times in a row. The precision from day to day was determined through measurement of four saliva samples and four spiked plasma samples (n = 8). The results are listed in Tables 4 and 5.

A relative standard deviation of over 10% is observed with plasma samples in a concentration range up to 2.5 mg/L. This is due to the resolution of the plate read-

Table 4 Within day precision for the determination of thiocyanate in plasma (n = 5–6) and saliva (n = 5–9).

Mean value [mg/L]	Standard deviation (rel.) s_w [%]	Prognostic range u [%]
Plasma		
12.41	4.2	10.8
6.70	2.0	5.1
2.49	11.4	29.3
1.33	17.4	48.4
Saliva		
167	2.9	8.1
120	2.2	6.1
44	2.7	6.2
38	1.4	3.9

Table 5 Day to day precision for the determination of thiocyanate in plasma (n = 8) and saliva (n = 8).

Mean value [mg/L]	Standard deviation (rel.) s_w [%]	Prognostic range u [%]
Plasma		
11.02	5.1	12.0
5.93	5.7	13.5
3.81	8.3	19.6
2.49	12.8	30.2
Saliva		
162	1.2	2.8
65	1.2	2.8
44	1.0	2.4
36	2.7	6.4

er of ± 0.001 extinction units (the measured extinction differences in this concentration range are between 0.003 and 0.009).

9.2 Accuracy

The accuracy of the method was checked by recovery experiments. Plasma and saliva samples were spiked with various amounts of thiocyanate and analysed. Results are listed in Table 6.

Table 6 Recovery rates for thiocyanate in plasma (n = 4) and saliva (n = 4).

Concentration without spiking [mg/L]	Analyte level after spiking [mg/L]	Recovery rate r [%]
Plasma		
0.627	12.08	98.4
0.627	6.34	99.0
0.802	2.71	114.8
Saliva		
36.60	112.7	100.0
54.61	92.94	104.5
19.75	38.92	103.0
36.60	44.15	100.0

9.3 Detection limits and quantitation limits

A detection limit of 0.76 mg/L was determined for the described procedure using the calibration graph method of DIN 32645 (German Industrial Standard [*Deutsche Industrie-Norm*]). This corresponds to a quantitation limit of 2.28 mg/L.

9.4 Sources of error

Excess thiocyanate concentrations may be observed in haemolytic samples and in the presence of acetoacetate in the sample [1]. Haemolysis of the samples must therefore be avoided. Acetoacetate concentrations with significant effect on the determination are rarely observed.

In plasma samples stored at –20°C for longer than 12 months, a precipitate is formed which cannot be centrifuged off and thus may affect the photometric determination adversely. Consequently analysis of the plasma samples should take place either directly after plasma collection or within two weeks, if the plasma is stored at –20°C.

10 Discussion of the method

The photometric method described here permits the quantitative determination of thiocyanate in small sample volumes and with a short sample preparation time. By adaption to microtiter plates, using a plate reader, the method is particularly suitable for the determination of large sample numbers. The thiocyanate levels in saliva are approx. 20 times higher than the corresponding levels in plasma (see Table 1). This is probably due to active secretion of the thiocyanate ion via the salivary glands. Furthermore thiocyanate concentrations in saliva strongly depend on salivary flow [15]. For this reason, thiocyanate levels in saliva are – despite of better precision – subjected to greater intra-individual variations than thiocyanate levels in plasma. Owing to the long half-life of thiocyanates, i.e. 6–14 days, this biomarker ought to be especially suitable for determining chronic exposures to low cyanide concentrations. For the given reasons, this at least applies to thiocyanate in saliva only to a limited extent.

Instruments used:

- Tecan GENios™ Multiplate reader with 492 nm filter, integrated plate shaking function and software Magellan 4.0 (Tecan, Germany).

11 References

1 P. Degiampietro, E. Peheim, D. Drew, H. Graf and J. P. Colombo: Determination of thiocyanate in plasma and saliva without deproteinisation and its validation as a smoking parameter. J. Clin. Chem. Clin. Biochem. 25, 711–717 (1987).
2 D. Henschler, J. Angerer and K. H. Schaller (eds.): Cyanide. Analyses of Hazardous Substances in Biological Materials: Methods for Biological Monitoring, Vol. 2, VCH, Weinheim (1988).
3 A. R. Pettigrew and G. S. Fell: Simplified colorimetric determination of thiocyanate in biological fluids, and its application to investigation of the toxic amblyopias. Clin. Chem. 18, 996–1000 (1972).
4 B. Junge: Changes in serum thiocyanate concentration on stopping smoking. Br. Med. J. 291, 22 (1985).
5 P. Eyer: Gasförmige Verbindungen. In: H. Marquardt and S. G. Schäfer (eds.): Lehrbuch der Toxikologie, 550. Wissenschaftsverlag, Mannheim (1994).
6 W. Steffens, D. Sibbing, N. Kiesselbach and J. Lewalter: Behandlung von Patienten mit Vergiftungen durch aliphatische oder olefinische Nitrile. Arbeitsmed. Sozialmed. Umweltmed. 33, 479–484 (1998).
7 M. Mueller and C. Borland: Delayed cyanide poisoning following acetonitrile ingestion. Postgrad. Med. J. 73, 299–300 (1997).
8 W. S. Rickert, J. C. Robinson, N. E. Collishaw and D. F. Bray: Estimating the hazards of "less hazardous" cigarettes. III. A study of the effect of various smoking conditions on yields of hydrogen cyanide and cigarette tar. J. Toxicol. Environ. Health 12, 39–54 (1983).
9 H. Greim, J. Angerer and K. H. Schaller (eds.): Cotinine. Analyses of Hazardous Substances in Biological Materials, Vol. 7. Wiley-VCH, Weinheim (2001).
10 R. G. Baumeister, H. Schievelbein and G. Zickgraf-Rüdel: Toxicological and clinical aspects of cyanide metabolism. Arzneimittelforschung 25, 1056–1064 (1975).
11 R. E. Bliss and K. A. O'Connell: Problems with thiocyanate as an index of smoking status: A critical review with suggestions for improving the usefulness of biochemical measures in smoking cessation research. Health Psychol. 3, 563–581 (1984).
12 J. K. Ockene, T. F. Pechacek, T. Vogt and K. Svendsen: Does switching from cigarettes to pipes or cigars reduce tobacco smoke exposure? Am. J. Public Health 77, 1412–1416 (1987).
13 K. J. Ruth and J. D. Neaton: Evaluation of two biological markers of tobacco exposure. Prev. Med. 20, 574–589 (1991).
14 M. Jarvis, H. Tunstall-Pedoe, C. Feyerabend, C. Vesey and Y. Salloojee: Biochemical markers of smoke absorption and self-reported exposure to passive smoking. J. Epidemiol. Community Health 38, 335–339 (1984).
15 D. Borgers and B. Junge: Thiocyanate as an indicator of tobacco smoking. Prev. Med. 8, 351–357 (1979).

Authors: *K. Riedel, H. W. Hagedorn, G. Scherer*
Examiner: *J. Angerer*

12 Appendix

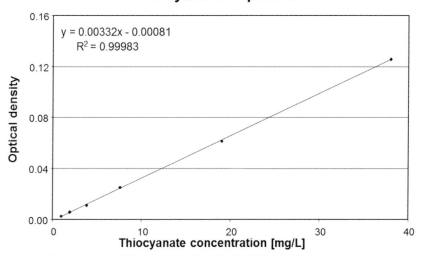

Fig. 3 Calibration graph of thiocyanate in plasma (optical density = absorption).

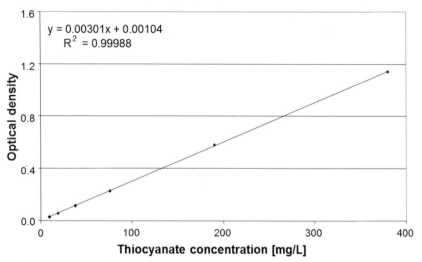

Fig. 4 Calibration graph for thiocyanate in saliva (optical density = absorption).

Members, Guests and ad-hoc Experts of the Working Group Analyses in Biological Materials of the Senate Commission of the Deutsche Forschungsgemeinschaft for the Investigation of Health Hazards of Chemical Compounds in the Work Area

Chair of the DFG Senate Commission	Prof. Dr. Andrea Hartwig Karlsruher Institut für Technologie (KIT) Institut für angewandte Biowissenschaften Abteilung Lebensmittelchemie und Toxikologie Adenauerring 20 D-76131 Karlsruhe
Chair of the Working Group	Prof. Dr. Thomas Göen Institut und Poliklinik für Arbeits-, Sozial- und Umweltmedizin Friedrich-Alexander-Universität Erlangen-Nürnberg Schillerstraße 25/29 D-91054 Erlangen
Members and permanent guests	Prof. Dr. Michael Bader BASF SE Occupational Medicine & Health Protection GUA/CB – H 306 D-67056 Ludwigshafen
	Dr. Meinolf Blaszkewicz ZE Analytische Chemie Leibniz-Institut für Arbeitsforschung an der TU Dortmund Ardeystraße 67 D-44139 Dortmund
	Dr. Ralph Hebisch Bundesanstalt für Arbeitsschutz und Arbeitsmedizin (BAuA) Leiter der Gruppe 4.4 – Gefahrstofflabor Friedrich-Henkel-Weg 1–25 D-44149 Dortmund
	Prof. Dr. Gabriele Leng Currenta GmbH & Co. OHG SI-GS-Biomonitoring Gebäude L9 D-51368 Leverkusen

Prof. Dr. Bernhard Michalke
Helmholtz Zentrum München
German Research Center for Environmental Health
Ingolstädter Landstrasse 1
D-85764 Neuherberg

PD Dr. Michael Müller
Institut für Arbeits- und Sozialmedizin
Georg-August-Universität Göttingen
Waldweg 37
D-37073 Göttingen

Guests

Dr. Edith Berger-Preiss
Fraunhofer Institut für Toxikologie und Experimentelle Medizin (ITEM)
Arbeitsguppe: Bio- und Umweltanalytik
Nikolai-Fuchs-Str. 1
D-30625 Hannover

Prof. Dr. Hendrik Emons
European Commission – Joint Research Centre
Institute for Reference Materials and Measurements
Retieseweg 111
B-2440 Geel
Belgium

Dr. Jochen Hardt
Institut für Laboratoriumsmedizin, Mikrobiologie und Umwelthygiene
Klinikum Augsburg
Stenglinstr. 2
D-86156 Augsburg

Dr. Peter Heitland
Medizinisches Labor Bremen
Dr. Wittke, Dr. Gerritzen und Partner
Haferwende 12
D-28357 Bremen

Dr. Hans-Wolfgang Hoppe
Medizinisches Labor Bremen
Dr. Wittke, Dr. Gerritzen und Partner
Haferwende 12
D-28357 Bremen

Dr. Holger Koch
Institut für Prävention und Arbeitsmedizin
der Deutschen Gesetzlichen Unfallversicherung
Institut der Ruhr-Universität-Bochum (IPA)
Bürkle-de-la-Camp Platz 1
D-44789 Bochum

Dr. Bernd Roßbach
Institut für Arbeits-, Sozial- und Umweltmedizin
Johannes Gutenberg-Universität Mainz
Obere Zahlbacher Str. 67
D-55131 Mainz

Prof. Dr. Gabriele Sabbioni
Tulane University School of Public Health and Tropical Medicine
Department of Environmental Health Sciences
1440 Canal St., Suite. 2100
New Orleans, LA 70112-2704
USA

Prof. Dr. Gerhard Scherer
ABF Analytisch-biologisches Forschungslabor München
Goethestrasse 20
D-80336 München

Dr. Thomas Schettgen
Institut für Arbeits- und Sozialmedizin
RWTH Aachen
Pauwelsstraße 30
D-52074 Aachen

PD Dr. Wolfgang Völkel
Bayerisches Landesamt für Gesundheit und Lebensmittelsicherheit (LGL Bayern)
Pfarrstrasse 3
D-80538 München

Dr. Tobias Weiß
Institut für Prävention und Arbeitsmedizin
der Deutschen Gesetzlichen Unfallversicherung
Institut der Ruhr-Universität-Bochum (IPA)
Bürkle-de-la-Camp Platz 1
D-44789 Bochum

Dr. Wolfgang Will
Im Bügen 12
D-67281 Kirchheim

Ad-hoc experts

Dr. Dana Barr
Department of Environmental and Occupational Health
Rollins School of Public Health of Emory University
1518 Clifton Road, NE, Room 272
Atlanta, GA 30322
USA

PD Dr. Lygia Therese Budnik
Ordinariat und Zentralinstitut für Arbeitsmedizin
und Maritime Medizin (ZfAM)
Marckmannstr. 129 b, Haus 3
D-20539 Hamburg

Dr. John Cocker
Health & Safety Laboratory
Harpur Hill, Buxton SK17 9JN
UK

Dr. Peter Fürst
Chemisches und Veterinäruntersuchungsamt Münsterland-Emscher-Lippe (CVUA-MEL)
Joseph-König-Str. 40
D-48147 Münster

Wolfgang Gries
Currenta GmbH & Co.OHG
SI-GS-Biomonitoring
Gebäude L9
D-51368 Leverkusen

Prof. Dr. Olf Herbarth
Umweltmedizin & Hygiene
Medizinische Fakultät der Universität Leipzig
Liebigstr. 27
D-04103 Leipzig

Dr. Olaf Päpke
Eurofins GfA Lab Service GmbH
Neuländer Kamp 1
D-21079 Hamburg

Dr. Roland Paul
Bundesanstalt für Arbeitsschutz und Arbeitsmedizin
Gruppe 4.2 – Biomarker
Nöldnerstraße 40/42
D-10317 Berlin

Jan Van Pul
Haven 725
Scheldelaan 600
B-2040 Antwerpen 4
Belgium

Scientific secretariat

Dr. Elisabeth Eckert
Institut und Poliklinik für Arbeits-, Sozial- und Umweltmedizin
Friedrich-Alexander-Universität Erlangen-Nürnberg
Schillerstrasse 25/29
D-91054 Erlangen

Dr. Anja Schäferhenrich
Institut und Poliklinik für Arbeits-, Sozial- und Umweltmedizin
Friedrich-Alexander-Universität Erlangen-Nürnberg
Schillerstrasse 25/29
D-91054 Erlangen

Secretariat of the Commission

Dr. Rudolf Schwabe
Kommissionssekretariat der Senatskommission der DFG
zur Prüfung gesundheitsschädlicher Arbeitsstoffe
Hohenbachernstraße 15–17
D-85350 Freising-Weihenstephan

Former members

PD Dr. Dr. Udo Knecht
Institut und Poliklinik für Arbeits- und Sozialmedizin
Aulweg 129
D-35392 Gießen

Prof. Dr. Peter Schramel
Saumweg 4
Ortsteil Bachenhausen
D-85777 Fahrenzhausen